BEHAVIOUR OF WOOD PRODUCTS IN FIRE

AF699086

WITHDRAWN

BEHAVIOUR OF WOOD PRODUCTS IN FIRE

Proceedings of a Seminar organized by the Timber Committee of the United Nations Economic Commission for Europe

Oxford, 22-25 March 1977

Published by
PERGAMON PRESS
for the
UNITED NATIONS

TH
1073
.B44
1977

U.K.	Pergamon Press Ltd., Headington Hill Hall, Oxford OX3 0BW, England
U.S.A.	Pergamon Press Inc., Maxwell House, Fairview Park, Elmsford, New York 10523, U.S.A.
CANADA	Pergamon of Canada Ltd., 75 The East Mall, Toronto, Ontario, Canada
AUSTRALIA	Pergamon Press (Aust.) Pty. Ltd., 19a Boundary Street, Rushcutters Bay, N.S.W. 2011, Australia
FRANCE	Pergamon Press SARL, 24 rue des Ecoles, 75240 Paris, Cedex 05, France
WEST GERMANY	Pergamon Press GmbH, 6242 Kronberg-Taunus, Pferdstrasse 1, West Germany

Copyright © 1977 United Nations

All Rights Reserved. No part of this publication may be reproduced, stored in a retrieval system or transmitted in any form or by any means: electronic, electrostatic, magnetic tape, mechanical, photocopying, recording or otherwise, without permission in writing from the publishers

First edition 1977

Library of Congress Cataloging in Publication Data
Main entry under title:

Behaviour of wood products in fire.

1. Fireproofing of wood--Congresses. 2. Building,
Wooden--Congresses. 3. Building laws--Congresses.
I. United Nations. Economic Commission of Europe.
Timber Committee.
TH1073.B44 1977 691'.1 77-24878
ISBN 0-08-021990-X pbk.

In order to make this volume available as economically and rapidly as possible the authors' typescripts have been reproduced in their original form. This method unfortunately has its typographical limitations but it is hoped that they in no way distract the reader.

Printed in Great Britain by
Page Bros. (Norwich) Ltd.

Contents

Introduction

Wood products constitute one of the most important groups of materials used in construction. No other construction material is so prominent in international trade.

Although wood is associated in many countries with the fire risk, attitudes are inconsistent. There are in fact wide differences in national building codes and regulations and in other factors governing the use of wood products in building.

The Timber Committee of the United Nations Economic Commission for Europe (ECE) noted in 1975 that governments were interested in the behaviour of wood products in fire. A seminar it had jointly sponsored earlier in that year with the ECE Committee on Housing, Building and Planning, at Sandefjord in Norway, on the industrial production and use of wood-based products in the building industry, emphasised the importance of the subject.

The Timber Committee therefore decided to hold a seminar on the behaviour of wood products in fire and to accept the offer of the United Kingdom Government to act as host.

The first aim of the seminar, it was decided, would be to reduce obstacles to international trade in wood products arising from inconsistencies in attitudes and differences in codes and regulations, mainly through efforts to harmonise national building regulations. The second purpose would be to increase the acceptance by the public, governments, drafters of regulations and financial houses, of wood products as a safe and rational construction material when used in the proper way and under suitable conditions.

The seminar would review research and practical experience in the behaviour of wood products in fire. It would relate such research and experience to building regulations concerning fire safety in force in ECE countries, and to other associations of the use of wood products with fire risks, as for example, in insurance and the setting of loan conditions. The seminar would also study the need for further action on the problem, and formulate recommendations to governments and international bodies, including the Timber Committee, as well as to the suppliers and users of wood products.

The seminar was held at St Peter's College, Oxford, from 22 to 25 March 1977. Some 60 representatives of 16 ECE member countries* and 8 inter-governmental and international non-governmental organisations took part.

---

*Austria, Belgium, Canada, Czechoslovakia, Denmark, Finland, France, Federal Republic of Germany, Ireland, Netherlands, Norway, Poland, Sweden, Switzerland, USSR, United Kingdom.

John Sunley (United Kingdom) and Yuri Tupitsyn (USSR) were elected, respectively, to the offices of chairman and vice-chairman. Felix Palmer (United Kingdom), chairman of the Timber Committee, declared the seminar open.

The seminar began with general discussions on fire and safety questions, building regulations and control, and particular problems of timber and wood products in fire. Papers were presented by experts from the United Kingdom and from the ECE secretariat. A list of national and international organisations concerned with the behaviour of wood products in fire, distributed to the participants, is included in these proceedings. The rapporteur for these discussions was Terence Mallinson (United Kingdom).

For the discussions on fire retardancy and fire resistance, papers were presented by experts from Canada, France, Poland, Sweden, the USSR and the United Kingdom. The rapporteur was Kurt Bengtson (Sweden).

As a final topic, the participants discussed moves towards the harmonisation of building regulations concerning fire safety and other fire-linked factors affecting the use of wood products. Papers were introduced by experts from Austria, the Federal Republic of Germany, Finland, the Netherlands, Poland, Sweden, Switzerland, the United Kingdom, the United States, the Comité européen des Assurances, and the ECE secretariat. Gérard Blachère, an ECE consultant, served as rapporteur.

Fire test methods in use and being developed in the United Kingdom were demonstrated for the benefit of participants in visits to the Timber Research and Development Association (TRADA) and to the Fire Research Station of the Building Research Establishment.

List of Contributors

Mr. H. Bednar, Senior Research Officer
University of Agriculture and Forestry
Gregor Mendelstr. 33, 1180 VIENNA
Austria

Mr. V. Biriukov, Director
UNIIDREV
Wood Research Institute
MOSCOW, USSR

Mr. B. Compin
IRABOIS
10 rue du Débarcadère
7582 PARIS CEDEX, France

Mr. P. Frech
Otto-Graf-Institut
FMPA Bauwesen, Pfaffenwaldring 4
7000 STUTTGART
Federal Republic of Germany

Mr. G.S. Hall, Head of Technology
TRADA, Stocking Lane
Hughenden Valley, HIGH WYCOMBE
Bucks HP14 4ND, United Kingdom

Mr. M. Heusse, Directeur
Développement "Produits Procédés"
ETS Rougier & Fils
75, avenue des Champs Elysées
75008 PARIS, France

Mr. H.L. Malhotra, Head
Building and Structures Division
Building Research Establishment
Fire Research Station
BOREHAMWOOD, Herts WD6 2BL
United Kingdom

Mr. A. Mårtenson
Head of Fire Protection Division
National Swedish Board of Physical
Planning and Building
Fack, 104 22 STOCKHOLM
Sweden

Mr. E. Mononen, Chief Engineer
Ministry of the Interior
Et Esplanadi 10
00130 HELSINKI 13, Finland

Ms Birgit Ostman & Mr. E.L. Back
Swedish Forest Products Research Laboratory
Box 5604, 114-86 STOCKHOLM
Sweden

Mr. J-C. Roux
Secrétaire général adjoint de l'Assemblée Plénière des Sociétés d'Assurances contre l'Incendie
11 rue Pillet-Will
75009 PARIS, France

Dr. J. Sunley, Director
TRADA, Stocking Lane
Hughenden Valley, HIGH WYCOMBE
Bucks HP14 4ND, United Kingdom

Mr. C. Ross Thomson, Director
Fire Research, Canadian Wood Council
Room 701, 170 Laurier Avenue West
OTTAWA, Ontario K1P 5Y5
Canada

Mr. Y. Tupitsin, Deputy Head of Department
MINLESPRON, Gritserets 2/16
MOSCOW, USSR

Mr. E. Urbanik, Head of Department
Wood Technology Institute
ul. Winiarska 1
60-654 POZNAN, Poland

Mr. T. Vilpuula, Managing Director
Puuinformaatio ry, Fabiankatu 29 C
00100 HELSINKI 10, Finland

Mr. R. Watson
Technical Director
Hickson's Timber Products Limited
Ings Lane, CASTLEFORD
West Yorkshire WF10 2JT
United Kingdom

General Aspects of Fire

Fire and Safety

H.L. Malhotra BSc (Eng), MICE, FI Fire E
Building Research Establishment, Fire Research Station,
United Kingdom

INTRODUCTION

The destructive nature of the 'unfriendly' fire has become a serious problem for many nations, particularly those with highly developed industries, congested urban population and having a high energy utilization in domestic and public environment. It might seem as if the complexity of modern living and industrial advancement went hand in hand with a high level of fire damage. A look at the latest data from a number of European countries shows the stark and gloomy picture.

TABLE 1 European Fire Losses, 1974

Country	No. of fires		Fire casualties			Estimated fire loss (per caput)
	Total	Per 10^6 capita	Total deaths	Per 10^6 capita	Injuries per 10^6 capita	
Austria	9 300	1240		N A	N A	6
Belgium	138 000	1410	160	16	74	N A
France	98 000	1870	310	6	40	12
Germany, Fed. Rep. of	N A	N A	N A	N A	N A	10
Italy	85 700	1550	220	4	130	3
Netherlands	22 700	1680	50	4	36	8
Norway		N A	70	17	N A	15
Sweden		N A	100	13	N A	13
Switzerland	9 400	1470	N A	N A	N A	7
United Kingdom a/	313 000	5600	1 000	18	116	9

a/ excluding Northern Ireland.

N A : Data not available.

Most fire incidents represent a misuse of a source of heat or energy by carelessness or in ignorance. Often the beginning of a fire is noticed and correct immediate action can control the situation. It has been suggested that the real number of incidents is perhaps much larger than indicated by the published data as a great many do not go beyond the primary ignition phase and do not become part of the fire statistics.

The awareness of fire has become an important part of the fire protection philosophy, awareness of its likelihood, its hostile nature when uncontrolled, the need for constant surveillance when certain potentially hazardous processes are undertaken and, perhaps most important of all, an awareness of the risk associated with some common everyday actions.

Has a certain extent of fire damage become a normal part of the civilized life in the same way as casualties in road accidents? It is neither theoretically possible nor economically feasible to be completely safe, but does the present situation represent a conscious decision or has society involuntarily allowed this level of fire damage to develop? This paper attempts to provide some notions on which fire safety decisions can be based.

WHAT IS A FIRE?

A fire can be characterized chemically as an exothermic process of oxidation and combustion, and physically by a rise in temperature of the ambient atmosphere and its pollution with fire gases. At ambient temperatures two important components of a fire, fuel and oxygen, exist in harmony ('R' Fig. 1) until the third component, a heat source, arrives on the scene. This causes a local instability, ('I'), due to a rise in temperature caused by the production of heat. If the temperature rises above 'I', heat release exceeds heat loss and point 'F', representing a dynamically stable condition of a fire, will be reached without further help and the fire will continue to burn until the fuel is exhausted or some fire control measures put in operation. On the other hand, if the temperature drops below 'I', cooling will take place with the gradual reduction in the size of the fire until the stable condition 'R' is reached again. Between the ambient stability at 'R' and the dynamic stability at 'F' there is a large zone of unstable situations which respond to changes in fuel supply, oxygen availability and the heat balance for the given environmental situation. Heating rates corresponding to less than 'a' will not cause a fire to occur, whereas greater than 'c' could lead to a flash fire. Heating rate 'b' represents a critical situation where transition from a stable to unstable condition is possible.

The beginning of most building fires is a modest heat source which comes in contact with a combustible material. At this stage the oxygen supply is adequate and the initial ignition can occur if the material temperature is raised sufficiently for combustion in the solid or the vapour phase. The continuation of the process is dependent upon the heat balance being positive i.e. an excess of heat production in comparison with the heat losses. The heat generated is partly dispersed to the atmosphere thus raising its temperature and partly used to heat the adjacent part of the material not involved in combustion to its decomposition point. When sufficient heat is not available and the rate of decomposition falls below a critical level a process of self extinction takes place. Most attempts at making combustible materials safe relate to transforming their characteristics so that the heat balance will not become positive, by raising the ignition temperature, dispersing the heat away from the material, preventing the ignition of flammable vapours or by intentional cooling.

DIFFERENT PHASES OF A FIRE

An uncontrolled fire usually passes through a number of different phases from its start to its final extinction when all the combustible materials have been virtually consumed. Often the individual phases may merge into each other without a clear distinction particularly if the fire is growing very rapidly. However it is useful for an understanding of the fire problems to consider it as consisting of different phases or stages.

The primary ignition phase at the beginning of the incident is when an individual item of the contents, particularly a material, is accidentally brought in contact with an ignition source which may be a lighted match, a cigarette end or an electric spark. Its particular characteristic is its small size and fairly low heat output: often such a source only lasts for a short time. For the ignition of the victim material to occur, it has to be in close proximity, its surface layer needs to be not only combustible but able to produce flammable vapours quicly, possibly with a high surface/mass ratio. Frequently the favourable conditions do not exist and the heat source is exhausted before ignition takes place. In some cases this phase may extend over a long period when smouldering precedes the active flaming. The progress is rarely controlled by the availability of oxygen as in the majority of cases a plentiful supply of air is available for the limited reaction taking place. At the end of this phase (Figs. 2 and 3) the ambient temperature may have been raised only marginally and if the fire is going to proceed to the next phase, another material or product has to become involved.

In the second phase another item of the contents, a furnishing material or in some cases an exposed wall or ceiling surface becomes ignited. The progression of fire will depend upon the flammability characteristics of the material, the orientation in which it is being used or if it forms a composite product. The nature of the substrate and the thermal inertia characteristics ($\kappa\rho c$) will influence its continued combustion. Gradual extinction is still possible if there is insufficient heat available to raise the temperature of the adjacent uninvolved part of the material to a level at which flammable vapours are emitted. With a positive heat balance, burning will continue at a progressively faster rate, the ambient temperature will rise (Figs. 2 and 4) increasing the speed of the chemical reaction. However, if ventilation is not adequate there will be some drop in the oxygen level with a consequential slowing down of the rate of decomposition. The fire progression is usually irregular and a number of happenings such as the breaking of glazing in its advancement. It is still possible for self extinction to take place due to insufficiency of oxygen or the complete consumption of the item if the exposure for the nearest uninvolved item is not high enough to cause its ignition. As the ambient temperature has been raised, the initial ignitability of materials becomes comparatively easier.

At the beginning of the next phase, fire spreads to an adjacent item, perhaps substantial spread over a wall surface occurs and flames touch the ceiling where accumulation of hot gases has taken place. The ambient temperature is still not high enough to lead to spontaneous ignition of the combustible surfaces. Provided the ventilation is adequate fire will progress from item to item, flames and hot gases along the ceiling will subject floor surfaces and other contents to radiation. The rise in the ambient temperature is at a more rapid rate (Figs. 2 and 5) and the end of the phase is leading to the point of no return, after which self extinction is improbable. The next part of this phase is a rapid progression of the fire, temperatures rising to 500°C and beyond, most materials igniting by convection or radiation, large quantities of smoke being generated and emitted through windows and other openings and progression to the total involvement and the 'flashover' stage. Just before 'flashover' the air temperatures are around 600°C, the level of heat in the room or compartment is such that flammable gases are emitted by almost all exposed combustible materials. Often the rapid progression is of a 'flash' type. As soon as this happens, unless the compartment is over-ventilated the oxygen supply is not adequate for all combustion to take place within the compartment. Consequently unburnt gases are emitted which burn outside the windows on the facade of the building (Figs. 2 and 6).

The fourth phase may be termed as the steady combustion phase when the fire involves the whole of the compartment or the building and the combustion of the contents and combustible constructional materials proceeds at the rate dictated by their nature, ventilation conditions and the compartment characteristics. This continues until the materials are virtually consumed and the fire dies out.

FIRE TESTS

The ability to predict fire behaviour in a given set of circumstances needs knowledge which is only partially available at present. Many studies are being undertaken to understand fully the physical and chemical reactions which take place and the inter-relation of different factors which exert an influence on fire behaviour. One aspect of this work is to identify the contribution of materials and products to a fire and the effects of fire on materials, components and constructions. As modelling of fire situations is still in an experimental state some of the appropriate information is obtained by means of fire tests. Such tests have been in use for many years but only in the last decade or two have their numbers increased significantly. Since the beginning of the present decade their role has been increasingly examined and more recently their very nature is being scrutinized.

In a number of countries the standards bodies are attempting to develop a philosophy for fire tests and the International Organization for Standardization (ISO) is pursuing a similar activity with a view to ensuring that its technical committees prepare tests of the appropriate type with a clear understanding of their functions. It is obvious from these discussions that fire tests are intimately linked with fire safety but often their precise contribution is not fully understood. On their own, tests are unable to indicate safety or hazard in a given situation, and they should be regarded as forming part of the overall scheme (Fig. 7) and capable of providing information on material and product behaviour which can be utilized in a systematic approach to fire safety. They are often used in this context as a part of the control system, allowing a selection to be made of those materials and products which can provide the necessary level of safety in a given fire situation. Tests can also be used as part of a research programme to assess fire hazard.

Four types of tests have been recognized (Table 2), on the basis of their nature and the measurement they can make. Often one test may fulfil more than one purpose and could be considered to belong to two types. This categorization permits a more rational attitude to be taken to the capabilities of different tests and the optimum use that may be made of them in a safety philosophy.

TABLE 2 Types of Tests

Name	Measurement	Feature
Basic property test	Thermal properties Calorific value	Material properties
Combustion characteristic test	Ignitability flammability, smoke, toxic products	Product behaviour
Thermal behaviour	Flame spread, fire resistance	Sub-assembly and assembly behaviour
Large scale *ad hoc* test. (Natural or semi-natural)	Fire spread, fire integrity, fire damage	Part or full structures

Tests which perhaps cause more confusion are the large-scale *ad hoc* tests, of the natural or the semi-natural type, depending upon the use of real contents or a simulation of the fire load; these are regarded as being the nearest approach to a fire situation. In a way this is true but it needs to be recognized that these tests only reproduce one particular situation and consequently their validity is limited. They are best utilized as a part of a research investigation, because their use as a demonstration can lead to subjective judgements being made on the suitability or otherwise of a particular form of construction.

FIRE SAFETY

Fire safety is becoming of increasing concern to all societies which are faced with a continuously rising loss by fire of life and property. Each fire incident is potentially capable of causing life loss as well as damage to property - the factors of concern are not necessarily the same and the mechanics used to exercise control are often different in nature as well as origin. In the past very little was done in either direction as a result of individual incidents involving a house or an individual business. Only when an incident of the magnitude of the Great Fire of London (1666) occurred, did the authorities take some action. In this case some restrictions were placed on the use of wood and, perhaps more importantly, the concept of risk separation was started by requiring properties to be separated by substantial constructions to resist fire penetration, with foundations laid for the use of space separation. It has been suggested that the wide boulevards of European cities were a device to prevent conflagrations. It is perhaps true to say that the main concern was to prevent large parts of cities being devastated by fire rather than to safeguard individual properties.

In the 18th and 19th centuries the advent of insurance and the need to protect investment in industry had a profound effect on fire protection principles. It started with a provision for fighting fires in the insured buildings by trained staff and simple appliances. This graduated towards an a assessment of the risk on the basis of commercial and industrial processes and the provision of suitable buildings. There was encouragement for better standards of construction, the use of safe materials and safe operational procedures. Nevertheless one of the main criteria was to balance risk against costs and therefore even unsafe situations were acceptable provided a suitable premium was forthcoming.

The approach has been refined over the last 100 years or so and now the insurance requirements are well defined, capable of dealing with complex industrial situations, and correlate requirements for structural construction and fire extinguishment and detection. The insurance companies have many sponsored research activities and are not reluctant to consider the latest developments in the fire protection field, provided they have been proved to be commercially viable.

Life safety concepts have progressed along a different route and have relied almost entirely on the community or society setting down standards through one or the other agency at its command. The provisions for this purpose were introduced slowly and progressively under a health or a safety provision in the local code or regulation. The initial requirements were rarely for the safety of individuals but of large groups particularly if these were in an unfamiliar place i.e. other than their homes or work places. The theatre regulations of the late 19th century following a serious fire in Vienna are a good example of this type of action.

Developments in the requirements for life safety have followed two separate routes. The first is to ensure the safety of people in those environments where they gather for pleasure, recreation, shopping, etc. and usually have only a limited knowledge of the layout of the building. In big centres and at certain times this can mean that a large number of people are present who are unaware of and unconcerned with safety. A number of incidents over the years, e.g. the Cocoanut Grove Fire in Boston, USA in 1942, the l'Innovation department store fire in Brussels in 1967, St Laurent du Pont dance hall fire in France in 1970, the Summerland, Isle of Man, fire in 1973 are examples of the scale of tragedy that is possible in our society and of the public concern which follows such incidents. The other area of safety concerns situations where people live or work, the surroundings are familiar, the need for protection may be a bit more apparent and the risk is a voluntary one within one's own domain. However the main concern is with safeguards against the folly of one person endangering the life of another e.g. fire in one house spreading to the adjacent house, or fire in one floor of an office spreading to the next one. The tragic fire in a multi-storey office building in Sao Paulo in 1973 is an illustration of the risk involved with buildings where large numbers of people are present and some of the elementary fire protection rules are ignored.

The current rules governing fire safety include complex requirements which deal with the design of the buildings, their internal arrangement, materials used in construction, materials used for finishes, provision of safe routes for evacuation of a part or the complete building, fire detection, extinction and control by fixed systems and aids for fire-fighting. The rules are often very complex but seem to lack coherence and appear to have developed somewhat haphazardly in response to one or the other incident which created a strong reaction in the community.

There is a need for a continually evolving system, rationally conceived and capable of responding to the new situations as they develop. As an example two problems need to be considered seriously in a scheme of this type: developments over the last two decades have put on the market many new products, for the home, office and the places of assembly, in the field of furniture and furnishing. These possess characteristics different from those found in the traditional materials, and an evaluation needs to be made not only of the increases if any in the fire risk but also the acceptability of this risk in view of the many benefits of such products.

The other problem concerns the relationship between the safety needs for life protection and property protection and whether both should not be combined as a part of a common strategy for fire protection.

A RATIONAL APPROACH

A rational approach to fire safety needs to be based upon a systematic analysis of the hazard and with the aim of providing an integrated solution. It should be based on the precept that complete safety is only a theoretical notion and that the level of safety should have an element of cost-effectiveness in it. Figure 8 illustrates an example of a safety square in which the analysis of fire risk and fire severity are interlinked with the aims and cost-effectiveness of the solutions. Each of these factors covers a number of sub-factors, the most important of which are identified in Table 3.

TABLE 3 Components of a fire safety system

Main factors	Sub-factors
Analysis of fire risk	Fire statistics Numbers in use Probability of fire
Estimation of fire severity	Type of occupancy Igniting sources Contents Fire development pattern
Aims of protection	Life safety Property safety Conflagration avoidance
Cost-effectiveness	Contents Warning & extinction devices Fire fighting

The analysis of the fire risk depends upon the reliability of the fire statistics and the information on buildings of various types so that a reliable probability prediction can be made. The basic data are collected in different ways in various countries, the usual sources being fire brigades, insurance companies and fire protection agencies. The need for a common scheme or at least an agreed list of minimum information is apparent if comparisons are to be made between different countries.

The estimation of fire severity can be made partly on statistical data indicating the broad pattern and partly on experimental investigations on growth and development of fires in buildings. The latter is being pursued actively by a number of research institutes and a reasonable predictive capacity exists on the modelling of fully-developed fires. However the fire growth pattern is more complex and modelling techniques are not as yet available. The main importance of this work would be to indicate sensitive areas for occupants' safety during the early stages of a fire and special needs for dealing with the smoke and toxicity problems.

The aims of protection need to be clearly enunciated and if possible distinctions made amongst measures to protect life, to protect property and to prevent conflagration. It is unlikely that some measures will be needed for more than one reason as often the life safety and property safety problems are interlinked. The incentives are different but it is possible that in future, government agencies will concern themselves with both problems as property losses can represent a serious strain on the national economy. Cost-effectiveness can be a controversial issue, nevertheless the validity of all judgements needs to be examined on the relation between expenditure and the level of safety achieved, particularly when alternatives are being considered. It is desirable to know the relative merits in quantifiable terms of controlling the nature of contents and/or the materials of construction as representing the loading of a fire against the provision of fire detection systems, sprinkler installations and improvements in fire-fighting facilities. It will also be necessary to consider several factors which take account of the acceptability of a voluntary risk against an involuntary one. The beneficial effects of new products in everyday life should be weighed against any increase in risk and the public inability to react to statistically infrequent dangers needs to be borne in mind.

REFERENCES

Emmons, H.W., Fire and Fire Protection, Scient. Am. 1974, 231

Harlow, D.W., International fire losses, Fire Jl 1975 69 (6) 43.

ACKNOWLEDGEMENT

This paper is contributed by permission of The Director, Building Research Establishment. The Fire Research Station is The Joint Fire Research Organization of The Environment and The Fire Offices' Committee.

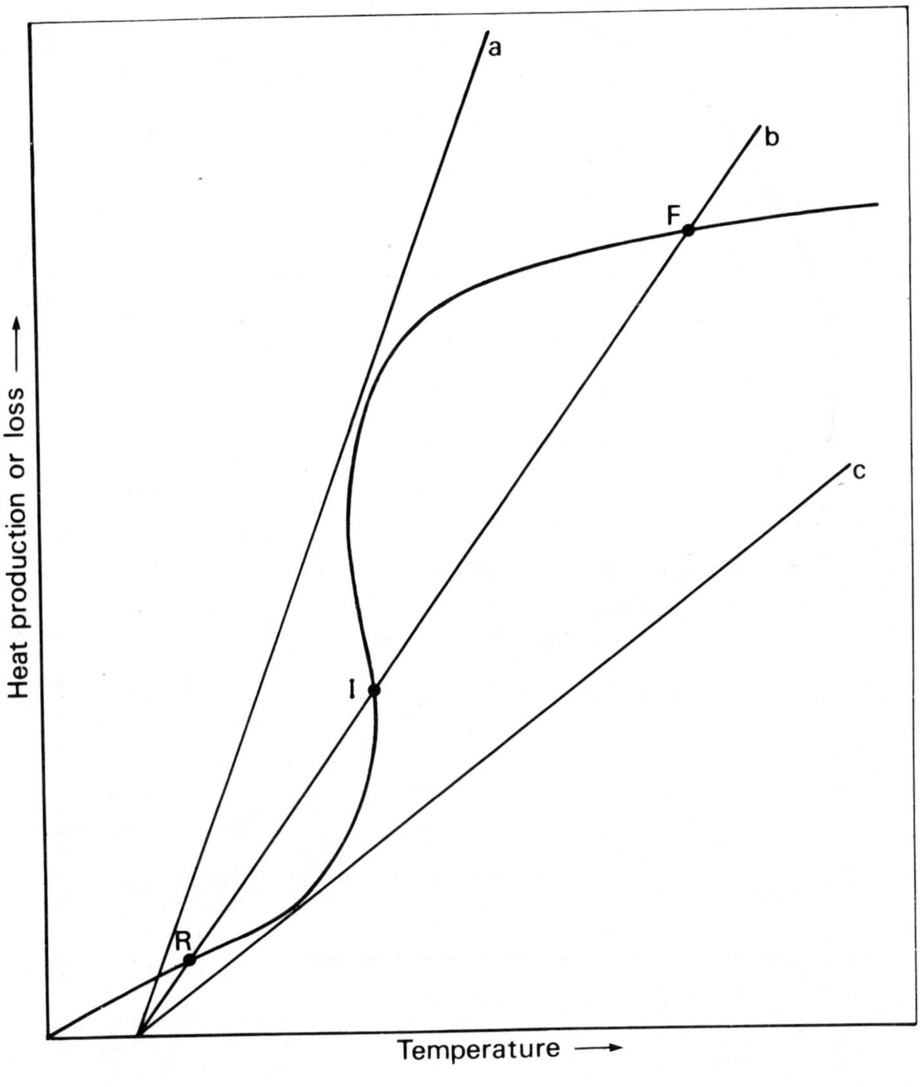

Figure 1 Fire dynamics

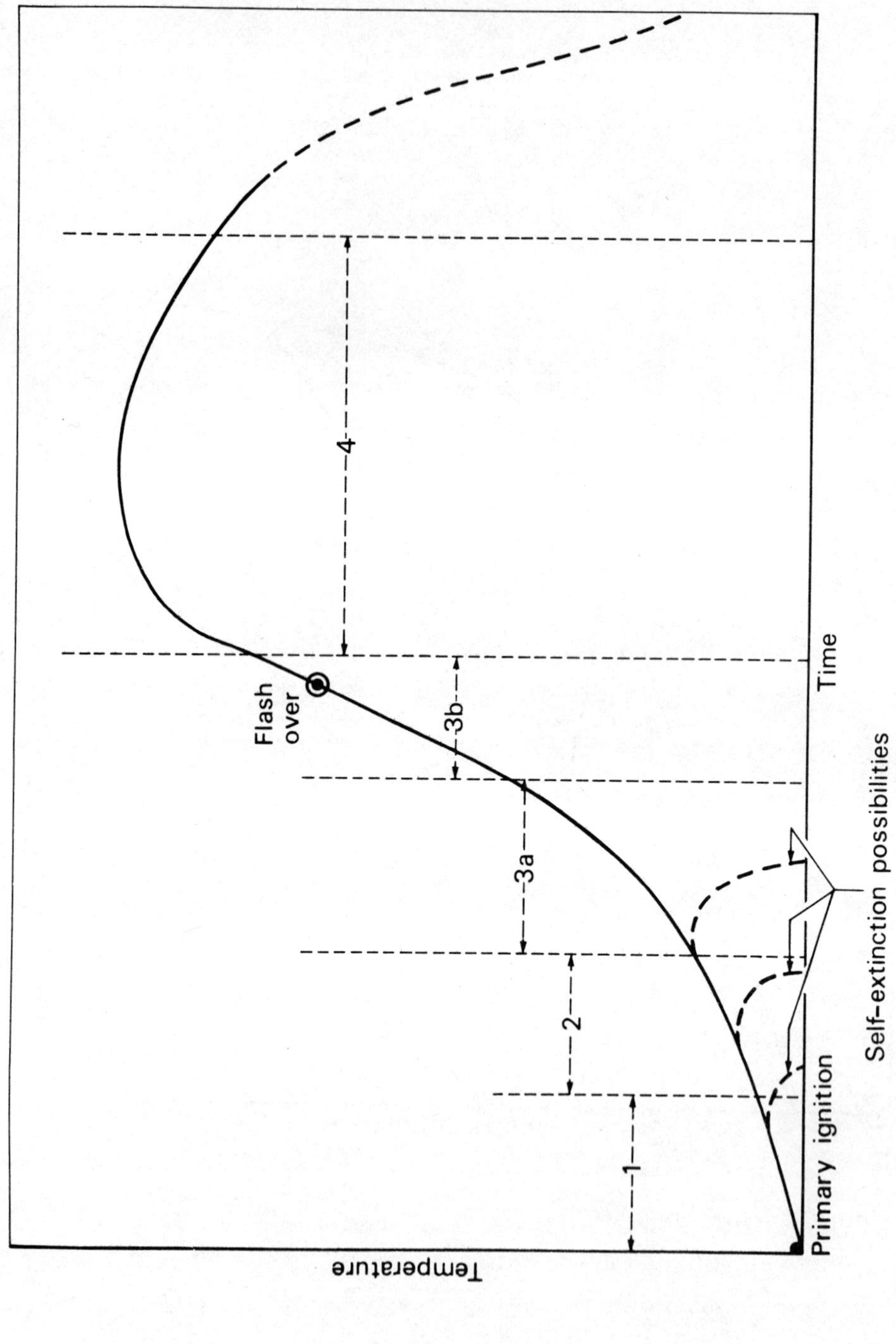

Figure 2 Different phases in the development of a fire

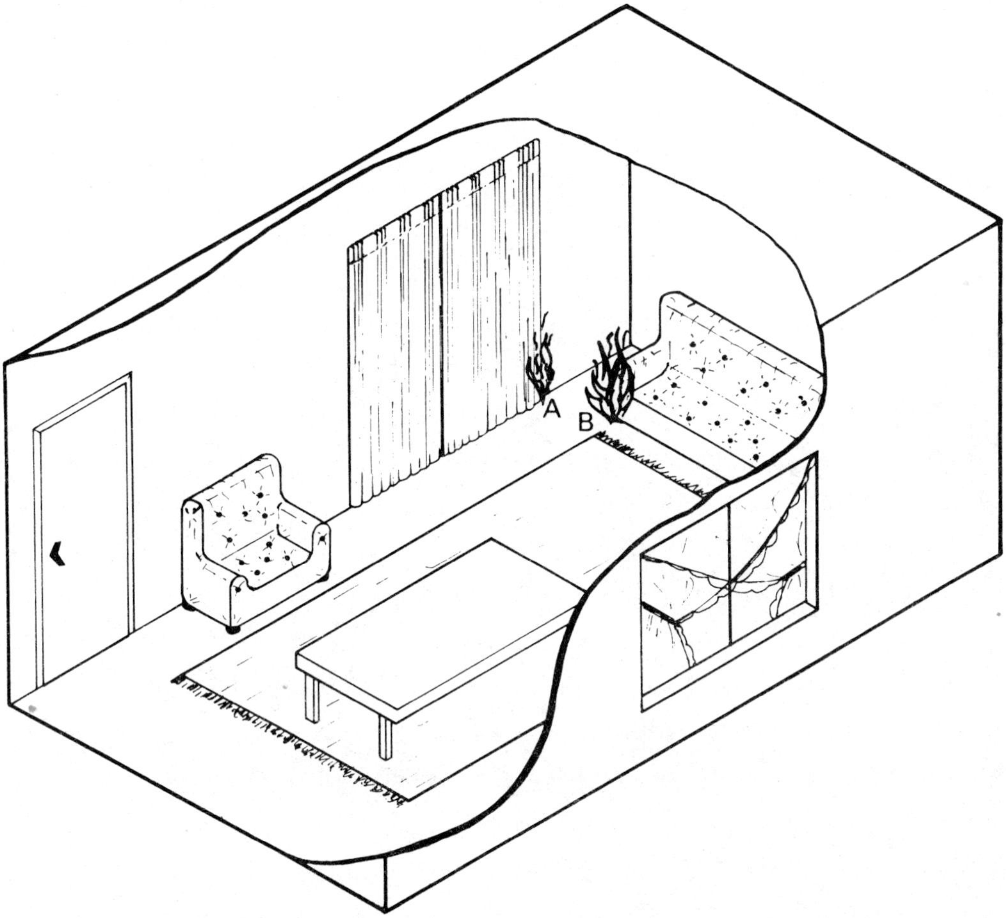

Figure 3 Start of a fire — primary ignition

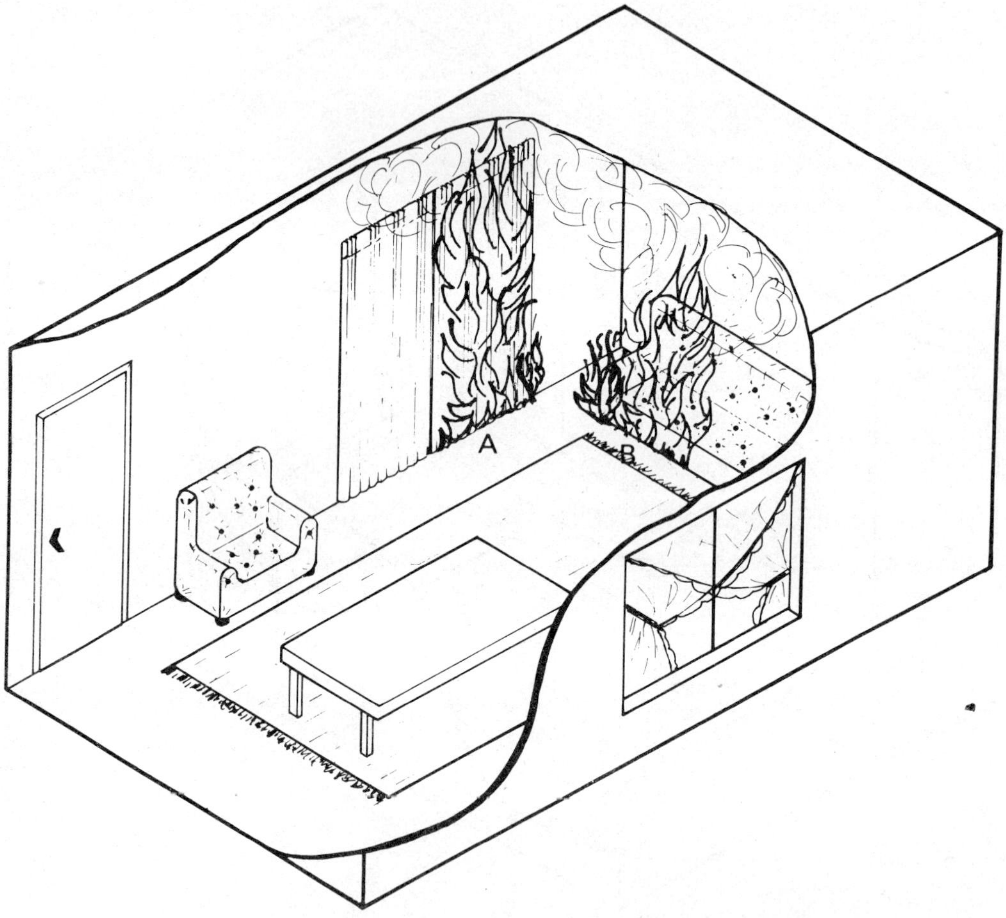

Figure 4 Development of a fire – initial item fully involved

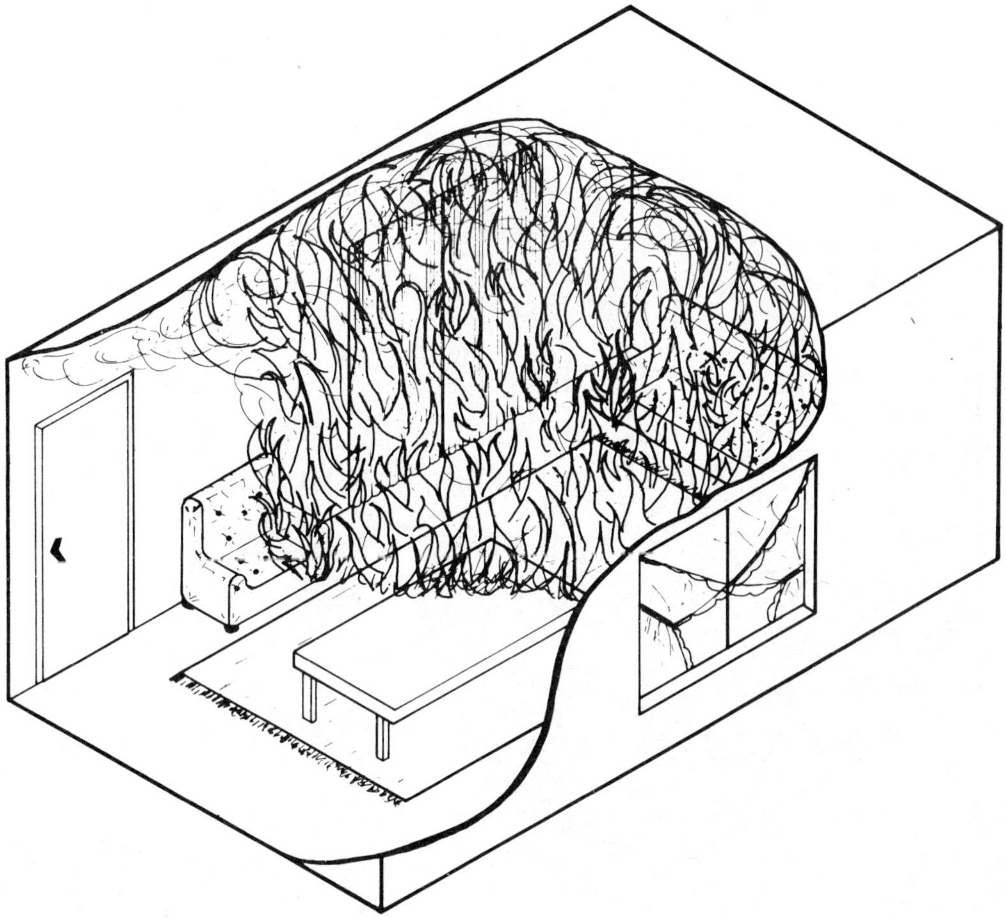

Figure 5 Growth of a fire — progressive involvement

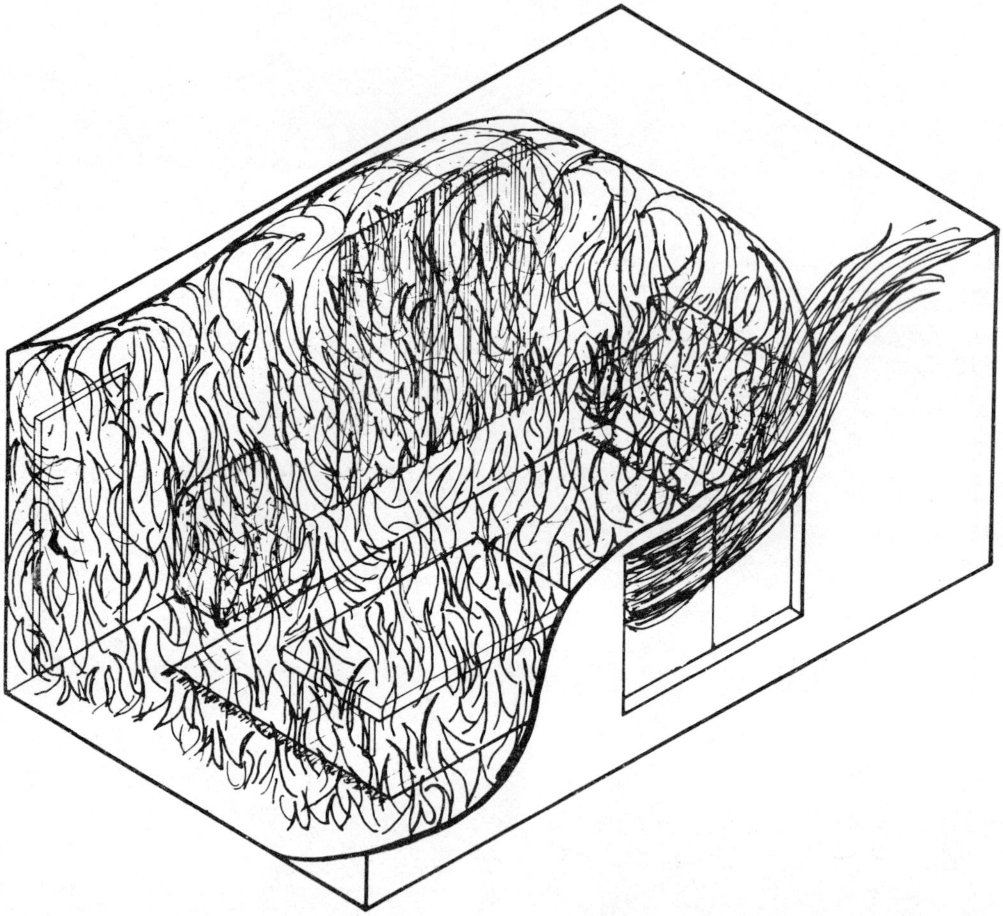

Figure 6 Fully grown fire

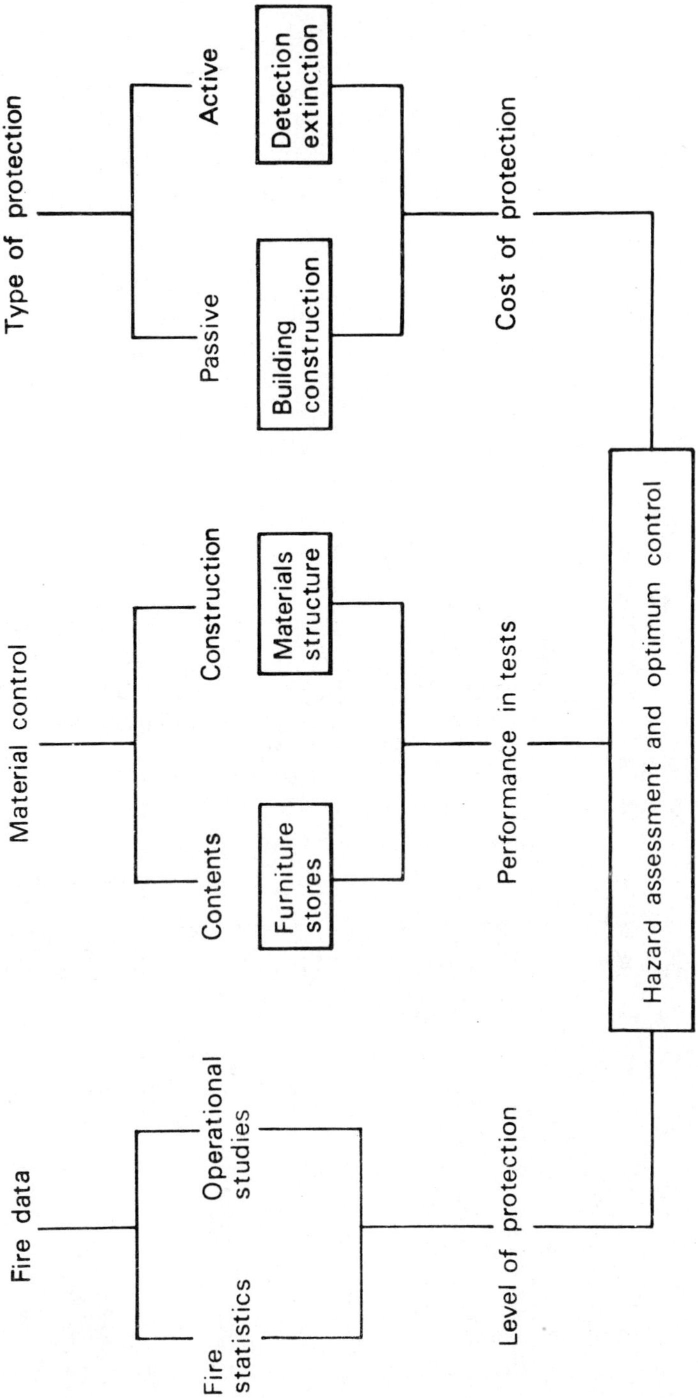

Figure 7 Fire protection strategy

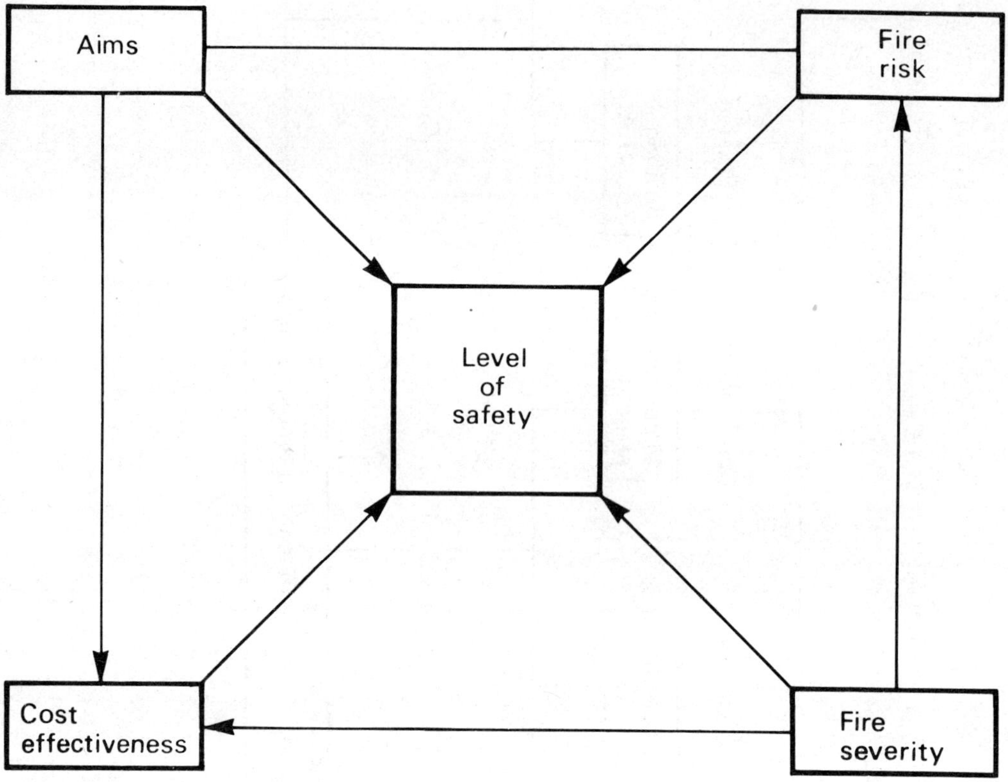

Figure 8 Fire safety square

Building Regulations and Control

Economic Commission for Europe, Environment and Human Settlements Division 1/

Building regulations are laid down to guarantee safety, health, and a minimum of comfort to the occupants. Those concerning comfort may vary considerably from one country to another, for they are closely dependent on the economic level of the country and its socio-cultural patterns. Those concerning health and safety are less determined by circumstances, for health and life are recognized everywhere as irreplaceable possessions though practice may sometimes differ widely. The concern that no human life should be lost as a result of fire is general.

But what are the regulations necessary to ensure safety and health? Must there be regulations at all?

Regulations have two aims. The first is to create a legal obligation to build well. Experience has shown that not all builders take the trouble, or have the capacity, to build well. The need is therefore felt to have a means of preventing the construction of unhealthy or dangerous buildings, or of bringing such buildings up to standard.

The second aim is to set forth the means of building safely, and then to see that they are practised, not only because some builders may be unaware of them, but also because, for many aspects of safety, nobody at all is sure what they are. In other words, knowledge is still incomplete. Merely to lay down that there shall be no more deaths from fire would simply mean placing the burden of a general lack of knowledge on a particular builder, should a tragedy occur. A closer look at the regulations thus shows them as a means of defending the builder. This is often overlooked in hasty judgements, but it is very quickly recognized if an offer is made to the builders to do away with all the details of the regulations in exchange for their acceptance of full responsibility.

Regulations are indispensable and must be as effective as possible. In other words, they must ensure health and safety at the lowest cost and with the least hindrance to innovation, and they must be easy to apply.

It is common knowledge today that the best rules are those which call for observance by defining the goals to be achieved, and not the means. Where fires are concerned, it would not be very useful for the builder if he were merely told repeatedly that no one should die in a fire. He would be unable to establish the link between this requirement and building specifications 2/. But it would perhaps be enough to require him to ensure that

---

1/ Mr. G. Blachère, consultant to the Secretariat, kindly co-operated in the preparation of this report

2/ On the other hand, in matters of acoustics, for example, it is quite sufficient to require that the sound level in a room shall be less than 40 dB.

flash over shall not occur in less than 15 minutes in the room where the fire has started, that the alarm shall be raised within five minutes of the fire's starting, that it shall be possible to evacuate the premises in less than 10 minites, and so on. Opinion is ready to accept regulations of this kind, but we need more knowledge if they are to be workable.

Meanwhile, it must suffice to state the goals to be achieved as the explanation of regulations which remain essentially descriptive of means - giving dimensions, masses, and the characteristics of reaction and resistance to fire.

The physical and chemical properties of materials do not yet allow us to predict their behaviour in fire, and so there is a need for tests simulating this behaviour in order to classify the materials. Until now such tests have been carried out only on small samples. But it happens that the size of the samples plays a role in their behaviour in fire. At present, therefore, there is a trend towards half-scale tests, at least as an intermediate stage until better tests on small samples can be worked out.

Each country has made some kind of test compulsory, placing an obligation on exporters to test their products several times. This is undeniably an obstacle to international trade.

Regulations, if they are to be effective, must not vary with the technologies or materials used (except where special precautions have to be taken because these are new or little known). Otherwise, they would create uneven conditions of competition. The materials cannot of course all have the same properties, but they must satisfy the same conditions. The ability to do this in respect of a given characteristic may be an inherent feature of one material, whereas the requirement may pose serious problems for another.

Safety regulations in any one country should form a homogeneous whole. This means that the risks incurred in the various daily activities should be the same - for example, the chance of dying should not be 2.7×10^{-4} per year in a road accident and 4.0×10^{-5} in a fire 1/ - and that the cost of saving one more life in a particular sector should not be disproportionate to that of doing so in another. Obviously the aim is to save as many lives as possible with the resources available. Thus, for example, the building of motorways is one of the least costly ways of saving human lives.

It is easy to say what effective regulations should do. It is incomparably more difficult to devise them, particularly since certain types of disaster and accident are fortunately exceptional and it is not simple to derive useful statistics from their study. What is more, it is always distressing to put one's name to a text which could mean the death of a number of human beings.

This is a matter, then, in which criticism is acceptable only if it also sets forth the remedy.

1/ Source: 1966 - United States Accidental Death Statistics.

International Organizations Concerned with Fire Problems

H.L. Malhotra BSc (Eng), MICE, FI Fire E
Building Research Establishment, Fire Research Station,
United Kingdom

One of the international organizations concerned with fire problems is the International Organization for Standardization (ISO). The main function of the ISO is to produce standards for products; however, it is also engaged on work leading to performance standards. One of its technical committees (TC/92) is concerned solely with the preparation of test methods to measure those characteristics of materials and products which have a bearing on fire behaviour. Its activities correspond with those of the National Standards Organizations but it is able to bring together a wider spectrum of knowledge and expertise in this area. It has a membership of thirty National Standards Organizations covering most of the European countries which also provide its most active component.

ISO operates without a sub-committee structure but with very active working groups, six of which are engaged on different groups of fire tests, one serves the instrumentation interests of others, another is studying the toxicity problems in a general way, one is active in the terminology field and finally a co-ordinating and policy making group comprising all the other chairmen and specialists. It has so far produced six international specifications and is busy on another ten or so at various stages of development. Its main output is test methods, i.e. fire test methodology for materials, components and construction which are found in buildings and which need to be assessed for their behavioural response under heat exposure conditions. This allows assessments to be made on aspects such as ignitability, flammability, flame spread, smoke production, and fire resistance.

A number of other technical committees of ISO also have an interest in fire tests from the viewpoint of particular materials e.g. TC 38, 45, 61, etc. These concern themselves mostly with a given material, rubber, plastics, textiles and the tests are aimed at examining their behaviour pattern as materials rather than as products in use.

The Central Secretariat of ISO became aware a few years ago of the need to co-ordinate activities on fire test development and has set up discussions and special panels to examine the problem and provide guidance. An _ad hoc_ panel, having a broad representation, under my chairmanship, has produced two documents, one dealing with the philosophy and basic concepts of fire tests and the other a guideline document for the technical committees. These are aimed at providing a unified and harmonized approach towards the development of fire tests and making them suitable instruments for use by the safety authorities. Fire tests are intended to provide information on which the safety authorities can decide whether the use of a particular material or product represents a hazard in a given situation.

The essential point from TC/92 to the seminar would be that in any fire hazard assessment tests fulfil a vital role and care is needed not only in their design but also in the use to which the information is put. An understanding of the role of fire tests and their application to practical situations is necessary as a real life situation may differ in many ways from the use of environment simulated in a test.

Timber and Wood Products in Fire — Particular Problems

G.S. Hall
Head of Technology, Timber Research and Development Association, United Kingdom

PROBLEMS OF ATTITUDE

More wood is used for fuel worldwide than for all other purposes put together and man used it this way long before he thought of using it for shelter, transport and many other of the innumerable uses to which it is now put. Because it is a renewable resource, he will probably continue to do so on an increasingly organized basis. This history of fuel usage is at the root of many of the problems afflicting the use of timber and its derivative building materials in fire resisting construction.

Because it is such a good fuel (when suitably prepared) its other attributes are either unappreciated or the attitude towards them is coloured by the race memory of camp fires, forest fires and sticks of kindling.

What is overlooked is that wood, to be a fuel, needs special preparation especially for getting fires started. This preparation, i.e. chopping into small pieces with a large surface area, is far removed from the way it is used as a structural building material which is, after all, the form in which we are most concerned when considering fire safety.

I believe that this tradition of fuel, which has been fundamental to the development of civilization and yet often barely under control, is, and is likely to remain, the most potent single obstruction in the optimum use of timber in buildings. People are afraid of fire, and rightly so. This applies as much to those of us who spend much of our time dealing with the subject as it does to the man in the street whose contact is via often lurid newspaper and television reports. Tell a man that timber studs lined on each side with 12.5 mm plasterboard will hold back a fully developed fire for half an hour and he will be unlikely to believe you. Perhaps the only way to convince him is to show him and demonstrate that near the end of a fire resistance test you can still put your hand comfortably on the outside of such a wall when the temperature in the furnace is approaching 1 000°C. Unfortunately, the opportunity to convince in this way arises very infrequently.

A similar misconception bedevils the flame spread properties of timber and wood-based sheet materials. Hardly a 'Western' goes by without instant 'flashover' of the structural timbers of the ranch-house or cabin occurring as a result of a flaming torch being waved in its general direction. Spectacular it may be, but wood just does not burn that way. Apart from veneer, wood wool and shavings, it is not easy to ignite and flamespread is not explosive.

Not surprisingly, the general attitude is that because wood is combustible it must be a hazard. Its use in buildings is rarely looked at rationally, particularly in those countries where brick or stone is the traditional building material. Probably for pragmatic reasons, a more reasonable attitude had evolved in those countries where timber construction is traditional. Certainly in the UK and probably in many neighbouring European countries, reports of serious fires often brand timber, plywood, chipboard or fibre building boards as culprits in circumstances which suggest their use as a scapegoat.

Thus the first particular problem associated with timber in fire situations is one of attitude. Deeply rooted in the subconscious and fostered by proponents of the so-called non-combustible materials, basic disbelief in the beneficial properties of timber in fire is widely prevalent from layman to building legislator.

PROBLEMS OF TECHNOLOGY

The second set of problems arises from the properties of the timber material and those others used with it. If the first problems are mental, then the second are physical.

We have to contend with natural variability of the material but then this is a characteristic which influences its use not only in fire situations. Processing, as in the manufacture of board materials, generally reduces variability. Solid timber is therefore most suspect from this point of view, but a fully developed fire as characterized by a fire resistance test seems to be a great leveller. For instance, natural defects, such as knots, char more slowly than adjoining clear timber. Differences between species seem to be submerged in a coarse relationship between density and fire behaviour.

Indeed the predictability of wood in fire is such that a method of calculating the fire resistance of a beam or column is possible and desirable. It enables structural fire protection with timber to be put on the same footing as other aspects of structural engineering. Such calculated values for fire resistance will, for the foreseeable future, have to stand alongside results of fire tests on prototype structures in terms of building codes and controls. It is therefore logical that both calculation and test should have the same parameters. The argument of predictability should not be taken too far out of context. It is perfectly feasible to produce rates of charring in timber which are far removed from 0.6 mm/minute by using high irradiance. Heat transfer by radiation may be greater in some fires than in some furnace tests. But prototype tests are carried out under the artificial fire condition represented by the requirements of ISO 834 and equivalent national fire resistance test standards. To retain comparability, calculation of fire resistance must also relate to the conditions of these tests and it is by these that standard charring rates have been measured and calculation procedures checked. It has been suggested that the fire resistance test may not give a very good representation of certain types of building fires. However, as a test it has perhaps the widest and longest history of use as a control of building materials and methods. Experience has not shown that the concept of the test method is deficient. The analysis of serious fires usually points up deficiencies in the way test results have been used or misused.

In the UK the British Standards Institution is currently involved in the drafting of a Code of Practice on the fire resistance performance of timber to supplement that on normal timber design. This code gives equivalent status to results obtained (i) by calculation, (ii) by fire resistance testing of prototypes and (iii) to a restricted range of constructions which have become established by long usage. For solid and glued-laminated beams and columns, the method of calculation involves the prediction of a notional residual cross-section, assumed to have prescribed mechanical properties, after exposure to standard fire condition has resulted in loss of timber by charring. The method is capable of considerable refinement, particularly in relation to joints and connections, on which work is currently being carried out. However, it is not appropriate to apply the method in this form to composite elements such as walls and floors. This is due in part to the complexity of the structure and in part to the behaviour of materials other than timber which react and interact differently under fire conditions.

Very rarely can a timber element be considered in isolation from other building materials. Even glulam beams and columns rely on the correct use of suitable adhesives for their fire resisting properties. The performance of plywood is particularly dependent on the type and quality of bonding. Nails, screws, staples, bolts, metal connectors, metal shoes, joist hangers and even door furniture can all be significant in determining the performance of the element as a whole. The number of potential combinations is enormous and the aim must be to establish performance on a generalized basis.

The approach being used to allow the calculation of fire resistance values for composite elements is merely a rationalization of what happens when confronted with a novel construction for test or when designing to achieve a successful test result. The construction is disassembled into a number of 'contributions to overall fire resistance'. These values are purely notional, but are based largely on a conservative estimate of performance derived from actual tests. The assigned contribution of, say, a wall lining will depend on the material, its thickness, the fixings used to attach it to the timber framework, the degree of support given by the framework, etc. These variables are entered as separate contributions where their influence is known or, more usually at this stage, constrained where their effect is not known or is in doubt.

Within basic rules, the contributions of exposed linings, loadbearing studs, cavity insulation and unexposed lining can be summed to arrive at a fire resistance rating for the construction as a whole. The figures are arbitrary and conservative but the method does allow for a limited amount of recombination to form constructions which have not been subjected to fire testing but whose performance can be judged on the basis of experience accumulated from testing similar constructions.

Similar alternatives to 'deemed-to-satisfy' tables have been used elsewhere, notably in Canada. In the UK Building Regulations, lists of acceptable type constructions are still the only alternative to the cost, delay and risk of prototype testing. Here at least then, timber, by virtue of its stability, high thermal insulation and known behaviour is pioneering a method which is far more flexible and less inequitable commercially than schedules of approved constructions. Being novel, there will obviously be problems in getting agreement on both principle and detail. Eventually, however, the method must be the way that codes of practice and associated building legislation express in detail the constructions deemed suitable for particular building situations.

Other systems become excessively cumbersome as more information emerges from continued testing. Tables of validated construction lengthen; indexing gets more difficult; physical limitations tempt reduction in detail with the consequent risk of errors in specification and quality control. With the predictive or calculation method, further testing will often result in greater refinement and reduction of constraints.

As an alternative to a prototype test with a test certificate, the methods being proposed in the code will undoubtedly meet with opposition as being less precise. Bear in mind, however, that a single test result is used to establish the likely performance of all subsequent constructions and the error of this assumption will be obvious.

Fire resistance of timber structures does present its problems but it can be demonstrated. Widely accepted tests are available to put results on a comparable basis and the properties of timber and commonly used adjunct materials are sufficiently well known to enable a predictive method to be put forward.

Reaction to fire is not in such a happy position. Test methods are divisive; there is lack of agreement on the significance of the various components of reaction to fire in various types of fire situation; the new generation of tests for ignitability, surface spread of flame and rate of heat release, which will give 'purer' results are slow in development and even then will require rigorous correlation exercises to establish levels of significance. National Building Regulations tend to be almost as individualistic as national standards. Both hinder trade and exchange of information.

PROBLEMS OF IGNORANCE

My third category of problems arises from what we do not yet know about fire and timber's behaviour i.e. problems arising from ignorance. Fire engineering is a relatively young branch of science and has a lot of ground to make up on longer established fields of engineering. Again there are two components of the problem, the one general, the other specifically related to timber.

One could go on at length about the merits of various test methods and their relevance to real or rationalized fire situations, but to little avail. What is important is that timber and its derivatives should not be unfairly discriminated against because the relevance of the test to actual fire safety is over-rated. In this connexion, the new ISO tests being developed should eventually be more capable of being justified and less suspect than our present ones. Their versatility will allow them to keep pace with improvement in our understanding of fire development and behaviour.

This problem besets not only timber and wood based sheet materials. It is one of getting standards that will be widely accepted and implemented and of drawing-up realistic building controls based on the relationship between such tests and real fire situations. These latter are so varied that only broad correspondence will be possible. What is more important in my opinion is that the tests are reasonably relevant to the feature that they will be used to control and are reproducible from laboratory to laboratory and country to country so that 'shopping around' for a satisfactory test result is prevented and international acceptance of test results is facilitated.

Timber linings, be they boards or sheet materials, are controlled under UK Building Regulations primarily on their surface spread of flame rating and to a lesser extent on results in the fire propegation test. This latter is a mixture of ignitability, flame spread and heat release. Based on a level which just accepts plasterboard, it is used to control linings used in escape routes, circulation areas in public buildings and other areas where the contribution of the room linings to fire growth must be minimal. The actual influence on fire development of a range of linings complying with this 'Class O' requirement has not been checked so that the limitation must still be considered arbitrary. As expected, its introduction has led to materials being developed to meet the test requirements. Their influence on fire development remains unchecked.

A similar situation exists in respect of flame spread ratings, although here there is a longer history of acceptance and implication that the levels set seem to work in practice. Here, problems with the use of timber or board products can be overcome by selecting products which have had their flame spread properties modified by the incorporation of flame retardant treatments. These treatments are not without their disadvantages and it is obviously inefficient to be forced into using them by regulations which might in practice not really be improving safety. This argument becomes very potent when the contents vs. structure issue is raised. Unduly onerous requirements for say, timber wall panelling, become very difficult to stomach when no controls of any kind are placed on furnishings and floor-to-ceiling curtains. However, this issue is wider than the problems affecting timber in fire situations, and this paper does not attempt to cover it.

An area of ignorance that at first sight should be simple to remedy but which is, nevertheless, one of doubt and disagreement, is that of fire statistics in relation to timber. Just how great is the risk of a structurally damaging fire in a timber frame dwelling compared with a traditionally constructed one? Is fire spread likely to be greater in one type than the other, given the same layout?

Here in the UK, a great deal of effort is put into collecting and analysing our fire statistics, and yet we cannot answer this type of question. No doubt the statistics are admirable for some of the uses to which they are put but when it comes to very specific aspects, reliable information is almost totally lacking. This in turn means that those who wish to promote the rational use of timber in building have no counter argument when certain mortgage or fire insurance interests claim to have actuarial evidence of a higher fire incidence or greater fire severity in this or that type of construction. Overseas experience, where timber frame construction is traditional, can be dismissed as irrelevant or misleading due to social and climatic differences. And so the argument goes on.

A soundly based analysis of fire risk and development would either prove what we have long suspected i.e. that timber frame houses are equivalent to (or better than) other types, or would point to the need for improvements in design, materials or workmanship. The chances of achieving such an objective are small. In a recent analysis of the UK fire statistics over a three-year period for a defined population of post-1965 Building Regulation dwellings, only 55 fires of <u>any</u> degree occurred in dwellings classed as timber framed. The difficulty of building up a pattern for structurally damaging fires on a strictly comparable basis with so low an incidence is obvious.

A problem for the future and probably the not too distant future, is that of smoke production. At present, timber tends to be the yardstick by which other materials are judged both in regard to obscuration and toxicity. Not all wood-based materials are in this position however, and those containing high proportions of organic binder or subjected to impregnation treatment are likely to be looked at most carefully. The question of smoke production and its effects embraces most of the problems besetting other aspects of fire performance. Questions such as -

Are the test methods reproducible?

Do they rank materials in an order that makes sense in real fire terms?

Are they consistent in their ranking?

What are realistic levels of control?

Is it possible to measure and control the important factors?

Is it worth controlling building materials at all when contents are uncontrolled?

And many more will have to be answered.

The answer in this case, as in other aspects of fire behaviour should be an objective balance between economics, safety and an attractive living environment. The objective of Building Regulations should be to achieve this and not to reflect traditions, pressures and reactions to isolated disasters.

Fire Retardancy and Fire Resistance

Fire Spread on Exposed and Concealed Surfaces and the Role of Fire Retardants

C.R. Thompson
Canadian Wood Council

> "The customary furnishings of most dwelling rooms provide enough fuel to create a serious fire regardless of whether the walls and ceilings are combustible or non combustible. Once a fire is started, it spreads not only by direct contact of the flames and by movement of hot air and combustible gases through channels not provided with fire stops, but also by elevating the surface temperature of combustible materials to their ignition point by direct radiation".

Thus concludes a 1959 report on "Experimental Dwelling-Room Fires" published by the US Department of Agriculture Forest Products Laboratory, reaffirming work done at Madison, Wisconsin in 1953. (1) In every country of the world fire takes its annual toll of life and property, and the report's conclusion sets the scenario for the course of a typical building fire - particularly those occurring in residential properties.

While present methods of gathering fire loss data (at least in North America) provide an incomplete statistical picture, some conclusions can be "deduced" from available data. The latest fire loss data for Canada (2) (ten year average, 1966-75) shows that 65% of the number of fires occur in residential properties but account for only 31% of the loss. The "average" loss per fire in that period was $1755 (the value of residential properties probably exceeds ten times that amount) indicating that the vast majority of accidental fires in these occupancies are controlled early or confined to the room of origin. Loss data reported for the Canadian Province of Quebec for 1974 (3) (to the author's knowledge, the most complete reporting done by any jurisdiction in North America) tends to confirm that fires in residential properties are kept small: 11,439 of 14,409 or 78% of residential property fires caused $1000 damage or less. Interestingly, the least number of residential fires in Quebec occurred at night, but as might be expected, the largest number of fire deaths occurred between the hours of 10:00 p.m. and 5:00 a.m., indicating that life loss occurs primarily when people are incapacitated, asleep or otherwise unable to take action to escape. The data does not, however, indicate any remarkable increase in dollar fire loss for "night" fires, leading to the conclusion that the building structure is seldom involved to any great extent and tending to confirm the commonly held notion that accidental fires in buildings usually involve the contents as the material first ignited.

Because most residential properties in North America (comprised of single family dwellings, row houses or small apartment buildings) are of timber frame construction, the interest of the forest products industry in fire research and conclusions like those reached in the USDA report is readily apparent; they help to correct popular misconceptions about the role of timber frame construction in fire spread.

This paper, therefore, deals with fire spread and will attempt to update the reader on some aspects as they affect timber frame construction.

The mechanism of fire spread (along the surface of a material, within the compartment of origin and outside it) are well documented.

Typically, fires in buildings involve the contents as the material first ignited from such causes as careless disposal of smoking material, faulty electrical wiring or heating appliances. The fire's progression and growth in the room of origin depends on many factors - principally the location (distribution) and combustibility of the contents, the ventilation available, the size and shape of the room and to some extent the thermal properties of the enclosing walls and ceiling. The fire is not usually influenced by the boundary elements of the room; it usually has sufficient fuel and air available in its initial stages. As the fire grows it involves increasingly larger amounts of combustible contents (either by continuity of flame travel or by radiation) - and it becomes more dependent on ventilation into the room. Usually, there is a sufficient influx of air to sustain combustion (through door and window openings); if not, and the fire has not been adequately ventilated to expel unburned combustible gases, flashover conditions can be quickly reached.

Flashover is usually described as the point in the development of a compartment fire when most exposed combustibles in the compartment become suddenly involved, dramatically aided by the unburned gases still unventilated from the compartment. Harmathy (4) indicates that the rate of fire development in the compartment is much more severe when the influx of air is just slightly below the point were a ventilation controlled fire is on the verge of becoming a fuel controlled fire, i.e. when the "control parameter", C, is very close to the critical value 0.079 $ft^{5/2}/lb$. 1/ In the experimental dwelling room fires conducted at the USDA Forest Products Laboratory flashover was reached in 4 to 7 minutes, an examination of the test report leads to the observation that the severity was due to the ventilation factor.

The part played by the room's contents cannot be minimized. In the typical dwelling room fire involving traditional types of furniture, the fire service's experience has been that flashover conditions are reached in a time frame of 5 to 20 minutes from ignition, the latter being common. With new materials used to simulate wood furniture (e.g. strippable foamed plastics) and expanded foam upholstery cushioning, time to flashover is being condensed into a 2 to 5 minute time frame, the former being common. In a demonstration test (5), an abandoned and vacant one storey wood frame bungalow having gypsum plaster on wood lath walls was furnished solely with a single baby's crib. The crib consisted of a rigid foamed plastic frame, metal spring and expanded plastic foam mattress. A single cotton sheet (the other other combustible in the room) was ignited and within 1½ minutes, flashover occurred in the room.

1/

$$C = \frac{A_W}{A_F} \cdot \frac{H_W^{\frac{1}{2}}}{F}$$

where A_W = area of windows, H_W = height of window, A_F = area of floor, F = fire load

Fire in a Provincial jail in Stratford, Ontario, on 18 December 1976, resulted in five prisoners losing their lives. The only combustibles were the five or six foam mattresses they piled in the corridor outside their cells and ingited by matches. Their plan was to feign being overcome by smoke and make their escape later from the prison's hospital. In duplicating the "scenario" the Ontario Fire Marshal's office found that the mattresses smouldered for about 2 to 3 minutes - then burst suddenly into flame releasing extreme temperatures and black acrid smoke. Observers to the tests had difficulty in making their escape under controlled conditions. The suddenness of the heat release accounts for the position in which one of the victims was found - on his back on the cell floor as if he was still holding on to the bars of the cell door.

The interior finish materials used to line the room of origin also play one or more parts in fire development, depending on their "degree" of combustibility:

- as the material first ignited
- as the medium providing the continuous surface over which flame can travel
- by acting as part of the fire resisting barrier (wall/ceiling/floor assembly) which helps contain the fire to the room of origin
- by acting as an insulating element to contain heat in the room (or conversely to act as a heat sink helping to dissipate heat from the room) and
- by contributing to the fire load - i.e. by releasing heat in the form of fuel into the fire.

If suppression action has not yet intervened to change the course of the fire, it has reached critical or flashover stage. It can quickly spread to other rooms and spaces, usually through opening communicating doorways, along hallways and up unprotected stairways to other floors. Heat and rising smoke and toxic gases make other areas quickly untenable to life and spread fire to other parts of the building, but as indicated from the statistical data, the inherent fire resistance of the gypsum board on timber frame boundary assemblies aided by active fire suppression usually confine fires to the room of origin.

Modern residential construction in North America is largely platform frame, where subflooring is continued out to the perimeter of the building and wood stud bearing walls are erected on the subflooring to carry the succeeding upper floors or roof. Concealed spaces between stud and gypsum board wall membranes which could act as vertical flues to carry fire from floor to floor are consequently fire stopped laterally by the stud and vertically by the floor and sill plates. It is unusual for fire to spread vertically in these fire stopped spaces. Fire stopping is place, however, where spaces concealed by interior finish fastened under stairs meets concealed floor joist spaces to reduce the risk of fire spreading undetected from floor to floor. In large buildings, it is usual to limit the area of concealed ceiling spaces to 3000 sq.ft., with no dimension exceeding 60 ft; and to place fire stops in concealed spaces over required fire separation walls or carry the separation up through the concealed space.

Fires that do spread in concealed spaces in residential construction usually occur in what the newspapers refer to as "substandard" housing; older buildings that have fallen into disrepair, lined with highly combustible interior finish, probably wired with the old "knob and tube" type of electric wiring, and most probably built of non-fire-stopped balloon frame construction. In this form of construction, the timber frame walls were built continuously for two or more storeys in height and the floor joists were framed into the walls such that the subflooring was not carried over the perimeter. The wall cavities were often not fire stopped at floor level, providing a flue for fire to travel from floor to floor.

For a more detailed discussion on fire spread, the reader is referred to the excellent literature survey published by the Timber Research Development Association in May 1972 (6). It contains descriptions of most fire test methods, an excellent discussion on the fire performance of timber as a material, and a review of how building elements using timber and wood base materials react in fire situations; Section 7 is a particularly valuable and extensive review of research and practical experience on the behaviour of wood products in fire, prepared by Gavin S. Hall. In his review Hall points out, quite rightly, that a typical fire is a rare occurrence because of the wide range of accidental fire situations and fire behaviour. Consequently, it may not be worthwhile to seek refinement in regulating or controlling the degree of combustibility of interior finish (lining) materials "when their involvement in fire is relatively rare and when their influence on the course of the fire is uncertain".

Unfortunately, building regulations do attempt to regulate interior finish materials on the basis of their "degree of combustibility". In North America they either wrongly use spread of flame test results (merely an indicator of relative surface flame propagation) or they outlaw "combustible" materials out of hand as a means to control potential fire severity. The rate at which a material will contribute fuel to a fire is a more logical indicator of its "degree of combustibility" for determining the necessary degree of control. The interest of the forest products industry in Heat Release Rate research programmes is therefore readily apparent.

The concept of Heat Release Rate Calorimetry (HRR) is the measurement of the rate at which a specimen releases heat energy when exposed to a controlled heat flux. Several HRR calorimeters have been developed (indeed, the 25 ft. Steiner flame spread tunnel is a form of HRR calorimeter because it provides data on the "fuel contributed" by a specimen) and five of these are described by David L. Chamberlain in a paper presented to the Combustion Institute in April 1975 (7). Chamberlain suggests that heat release rates are a more realistic and usable means to define and measure degree of combustibility than any other currently in use.

The forest products industry in North America, through the National Forest Products Association, has been sponsoring a fire research programme spearheaded by Chamberlain at the US National Bureau of Standards (NBS). This programme has three major thrusts:

- determining the Heat Release Rate performance of elementary wood products using the NBS Calorimeter.

- determining the Heat Release Rate performance of timber frame assemblies using the Factory Mutual (FM) Construction Materials Calorimeter, and

- determining the Heat Release Rate of full scale construction assemblies using the standard ASTM E-119 fire endurance test method as a calorimeter.

The purpose of the latter two objectives is to determine first, if it is possible to obtain Heat Release Rates on representative assemblies (and not just elementary materials) using the relatively small FM Calorimeter; secondly, if it is possible to use a "Standard E-119 furnace as a calorimeter to determine Heat Release Rates from large assemblies and thirdly, what relationship exists between the three - with a possible view to adding another dimension to the use of the apparatus involved.

NBS HEAT RELEASE RATE CALORIMETER (Figure 1)

The work on heat release rate measurements at NBS is nearing completion and a report on the heat release properties of several wood products is expected later this year.

The NBS HRR Calorimeter is capable of exposing material samples to a wide range (up to 10 watts/cm2) of radiant heat flux, enabling it to evaluate the response of materials to a wide range of fire conditions. It also has the advantages of small sample size (11.4 cm x 15.2 cm) and uniformity of exposure. Above exposures of about 4W/cm2, auto-ignition of most plastic and cellulosic materials occurs in 5 to 60 seconds; below that flux, pilot flame ignition is required. The sample holder is water cooled to remove heat from the back face of the sample; and the temperature rise in the water stream permits the calculation of heat released in that manner. The apparatus is designed to burn pyrolysis products completely by means of a propane (liquified petroleum gas) diffusion flame in the secondary burner. The change in propane flow to the secondary burner is a direct measure of the rate of heat release.

Some preliminary results of Chamberlain's experiments are shown in table 1 (8). For the experiments, exposure of 2.6, 4.0, 6.0 and 8.0 W/cm2 were chosen. These results indicate that, as expected for most wood based materials, heat release rates increase with increasing radiant flux. But, from 6.0 to 8.0 W/cm2 exposure, the rates for wood unexplainably decrease; the effect of this phenomenon on fire endurance will be explored. It was also found that fire retardant treatment (wood pressure impregnated with fire retardant chemicals to high retention levels) has a pronounced effect in reducing rates at which heat is released but has a minor effect upon times to ignition and peak rates, a significant factor to consider when assessing the use of fire retardant treated wood as an interior finish material.

Exposures of 2.6 W/cm2 are insufficient to ignite most wood and cellulosic products, and a pilot flame in two positions was used to help shed some light on the most appropriate pilot flame locations. The difference in results between the two positions is illustrated by the marked differences between time to ignition, time to peak rate, and peak heat release rate. Which of the two positions is the most "realistic" is at the moment a matter of conjecture, indicating that the test method itself must undergo considerable refinement before it can be used as a standard "test method" to evaluate a wide range of materials.

FM CONSTRUCTION MATERIALS CALORIMETER

The FM calorimeter (Figure 2) has been in use for a number of years to evaluate floor, roof and wall assemblies. The relatively large sample (121.9 cm x 121.9 cm) is exposed to a heat flux approximating the ASTM E-119 standard time temperature curve compressed in a shorter time. Maximum heat flux is about 15 W/cm2. Interior finish materials are exposed for 10 minutes, construction assemblies for 30 minutes. Flue gas temperature is recorded throughout the test; an inert (reference) sample is then substituted and given identical exposure. The rate of heat release is obtained from the rate of consumption of propane (liquified petroleum gas) which is burned to maintain the flue gas time-temperature curve during the substitution run.

In the National Forest Products Association tests, a 70 minute exposure following the ASTM E-119 standard time-temperature curve was used (9). A single model wall assembly, constructed of 2" x 4" lumber and sheathed on each side with 5/8" type x (a special type of fire resisting) gypsum board was tested in the FM calorimter to

(a) determine the heat release characteristics of a large "protected" timber frame sample, and

(b) to compare heat release performance of the model assembly in different calorimeters.

(Small sized wood frame and gypsum wall board samples have been run in the USDA/FPL Calorimeter and the Stanford Research Institute Calorimeter in California. Results of the USDA/FPL tests were reported by Brendon (10) in 1974).

The exposure was comparatively uneventful; temperature on the top surface of the specimen at 60 minutes was estimated at about 60°C. At 60 minutes, gypsum board adjacent to a seam began to sag but the joint was still intact; at 70 minutes when the test was concluded, the seam had opened for about 2-3 ft. in length and the interior of the stud cavity was burning vigorously. Extensive pyrolysis and charring of the wood occurred within the stud cavities; however, the shape and size of the original cross section was retained by the char leading to a tentative conclusion that scarcity of oxygen in the stud cavity prevented extensive combustion of either char or pyrolysis products within the wall cavities.

Heat release data are plotted in Figure 3. The maximum release rate of 2.8 W/cm2 occurred as a sharp peak at about 8 minutes - probably due to burning paper on the exposed gypsum board wall surface; the initial short peak at about 3 minutes was probably due to flame spreading over the paper wall facing. The cumulative heat release of these two peaks at 10 minutes is approximately 7800 British Thermal Units (BTU); heat of combustion of paper on 16 sq.ft. of gypsum board was calculated at 7830 BTU. Total heat available from the exposed wood in the wall was calculated at 171,000 BTU, including heat available from all paper surfaces of the gypsum board, total heat available from the assembly was estimated at 195,000 BTU.

From the heat release curve, the fraction of estimated total available heat that was released at a given time during the test may be calculated, e.g. at 60 minutes this 1-hr. fire resistance rated assembly had released only 25%

of its available heat; 4% of which came from the paper. Further, assuming that 17,700 BTU was released from all paper in the test assembly (6,900 BTU remaining in residual char) the total heat released by the wood at 60 minutes is 31,800 BTU, or only 16% of the total available.

It is also of interest to compare heat released from the assembly to the heat input of the furnace. Total heat input into the furnace was 815,000 BTU at 36 minutes and 1,745,000 BTU at 70 minutes, while total heat released from the test assembly was 24,300 BTU and 68,000 BTU respectively. Taking into account the heat released by the paper, the estimated heat released by the wood studs would represent 2.0% of the total heat input at 35 minutes, and 3.5% at 70 minutes.

The small amount of heat released into the furnace by the wood studs during the test period was reported to have surprised the FM staff conducting the tests.

ASTM E-119 FIRE ENDURANCE TEST (as a Heat Release Rate Calorimeter)

A mass flow meter was installed in the fuel line to a standard furnace used for fire endurance wall tests conducted under ASTM E-119 criteria (11). Fuel consumption rate required to maintain the desired furnace temperature when exposing the non combustible reference specimens (gypsum board on steel stud walls) is used as a "base line" and compared with fuel consumption for a typical timber framed assembly (gypsum board on wood stud wall). Any heat released or absorbed by the timber framed assembly will be balanced by an equal reduction of increase in fuel consumption required to maintain the predetermined (standard) time-temperature environment in the furnace. Heat release rate from the test (timber frame) specimen is obtained as the difference between the two consumption rates.

Fuel consumption data have been obtained on five wall assemblies that have potential use as "reference" assemblies (table 2) (12). The one timber frame/gypsum board assembly tested so far, had a fuel consumption rate almost the same, for 60 minutes, as the rates for the reference assemblies. From these limited data, it is clear that the heat release from an intact timber frame/gypsum board assembly is low, and does not become significant until the protective membrane is ruptured.

FIRE RETARDANT TREATED WOOD

Fire retardant treated wood is wood that has had its characteristic behaviour to fire exposure altered by pressure impregnation with fire retardant chemicals. How chemical impregnation reduced the "flammability" of wood was best summed up by Eickner of USDA/RPL in his August 20, 1975 letter to Buchan, copy of which is appended to this paper (Annex).

The HRR characteristics of fire retardant treated wood have already been discussed; additional calorimeter tests conducted by FM on a 1/2 in. thick 4½ ft x 5 ft specimen of fire retardant treated plywood showed a maximum one minute heat release rate of 161 BTU/sq ft/min, falling below the 200 BTU/sq ft/min value generally associated by FM with the most low fire hazard building materials not requiring automatic sprinkler protection in and of themselves.

To further evaluate fire retardant treated plywood, the American Wood Preservers' Institute conducted an exploratory fire hazard examination in the Factory Mutual full scale building corner test facility (13). Purpose of the test was to evaluate the burning characteristics of the plywood panels when subjected to an ignition source simulating a building fire environment in an otherwise non combustible occupancy. To be acceptable, the material being tested could not produce a self-propagating fire to the limits of the structure as evidenced by flaming or material damage.

Fire exposure in this 15 to 20 minute test is provided by a 5 ft. high pile of 4 ft x 4 ft wood pallets (weighing about 750 lbs) arranged to simulate the ASTM standard time-temperature curve for a minimum of 15 minutes and producing 1000°F in 5 minutes or less. The fire exposure pile is located at floor level in the building corner and 1 ft. away from each wall. Temperatures at various locations are monitored by thermocouples.

Extent of flame propagation and panel damage as shown in Figure 4. The results compared favourably with those obtained under similar test conditions with paper faced gypsum wall board and did not produce a self-propagating fire. It was interesting to note that maximum temperature at the ceiling in the corner reached 1415°F with gypsum board walls and only 1465° with the treated plywood, hardly a significant difference !

REFERENCES

(1) Bruce, H.D., April 1959. Experimental Dwelling-Room Fires, Report No. 1941, US Dept. of Agriculture, Forest Products Laboratory, Madison, Wisc. USA.

(2) Fire Losses. Canada, 1975; Report of the Dominion Fire Commissioner, Dec. 1 1976, Gov't of Canada, Department of Public Works, Ottawa, Ontario, Canada.

(3) Statistiques sur les pertes causées par l'incendie 1974 Statistics of fire losses; Report of the Director General Fire Prevention Branch, Ministry of Municipal Affairs, Gov't of Quebec, Quebec, Canada.

(4) Harmathy, T.Z. 1976, Fire Resistance versus Flame Spread Resistance, Fire Technology November 1976, National Fire Protection Association; Boston, Mass., USA.

(5) Society of Fire Protection Engineers, Committee on Fire Phenomena, 1975; Hazard Alert, and internal report on experiments at University of California at Berkley, SFPE Bulletin, April 1975, Society of Fire Protection Engineers, Boston, Mass., USA.

(6) G.S. Hall, R.G. Saunders, R.T. Allcorn, P.E. Jackman, M.W. Hickey and R. Fitt; Fire Performance of Timber, TRADA, a literature survey; Timber Research and Development Association, High Wycombe, United Kingdom.

(7) Chamberlain, David L., 1975, Rate of Heat Release - Tool for the Evaluation of the Fire Performance of Materials; paper presented to Combustion Institute at San Antonio, Texas, April 21-22 1975; National Forest Products Association, Washington, D.C., USA.

(8) Chamberlain, David L., 1976, Heat Release Rate Performance of Wood Products in NBS Calorimeter; November 4 1976, Internal Report to National Forest Products Association, Washington, D.C., USA.

(9) Chamberlain, David L., 1976, Heat Release of a Protected Wood Frame Model Wall Assembly in the FM Construction Materials Calorimeter; April 28 1976, Internal Report to National Forest Products Association, Washington, D.C., USA.

(10) Brendon, John J., 1974, Rate of Heat Release from Wood-Base Building Materials Exposed to Fire; FPL Research Paper No. 230, US Dept. of Agriculture, Forest Products Laboratory, Madison, Wisc., USA.

(11) National Forest Products Association, 1976, ASTM E-119-73 Fire Endurance Test, Load Bearing Partition Wood Studs with Gypsum Wallboard, Design WP 395; Serial 27306, January 8 1976, Factory Mutual Research, Norwood, Mass., USA.

(12) Chamberlain, David L., 1976, Heat Release Rate Measurement in ASTM Standard E119 Fire Endurance Tests on Full-Scale Construction Assemblies; October 30 1976, Internal Report to National Forest Products Association, Washington, D.C., USA.

(13) American Wood Preservers Institute; 1975, Exploratory Fire Hazard Examination, Full Scale Building Corner Test of Fire Retardant Treated Plywood Panels; Serial No. 24697, July 15, 1975, Factory Mutual Research, Norwood, Mass., USA.

Extracts of letter dated 20 August 1975 from Mr. H.W. Eickner, Research Project Leader, Improved Fire Performance of Wood, US Department of Agriculture, Forest Products Laboratory to Mr. R.B. Buchan, National Forest Products Association

"Mr. Eickner summarized the theories on how fire-retardant chemicals reduce the flammability of wood in the following way:

1. Nonflammable gases are released from the decomposition of the chemicals which dilute any combustible gases formed to result in a noncombustible mixture.

2. Thermal conductivity of the wood is increased, thus dissipating heat faster than can be supplied by igniting sources.

3. Chemicals undergo chemical and physical changes that absorb enough heat to reduce pyrolysis actions of wood surfaces and prevent ignition.

4. Fire-retardant chemicals form a liquid or glassy layer which prevents escape of flammable products and also prevents air from reaching wood surface.

5. Fire-retardant chemicals form foams of low density, either alone or in combination with wood surface, which insulates the wood subsurface and prevents the pyrolysis to flammable products.

6. Fire-retardant chemicals release certain active atoms at pyrolytic temperatures that restrict the ratio range over which pyrolysis gas-air mixtures will be flammable.

7. Fire-retardant chemicals decompose at temperatures slightly less than that for normal wood pyrolysis releasing strong acid or basic products. These chemicals promote the more complete decomposition of the wood to charcoal, carbon dioxide, and water instead of the formation intermediate decomposition products such as flammable tars and gases responsible for the flaming combustion.

It is likely that the action of fire-retardants chemicals is obtained by processes represented by several of these theories. However, recent studies have generally indicated that the last theory "controlled pyrolysis to form more charcoal, carbon dioxide, and water and less flammable tars" is the most valid.

Fire-retardant chemicals including certain phosphates and boric acid also control the burning of wood by preventing the afterglow of the charcoal residue. The theory here is that the chemicals control the initial oxidation process so that the molar ratio of carbon monoxide to carbon dioxide is high. This results in much less exothermic heat release, and therefore the oxidation process does not continue. As a result the total carbon monoxide-carbon dioxide products formed are reduced.

TABLE 1: Average Heat Release Properties of Wood Products In NBS Calorimeter

(Conditioned At 50% R.H.)

Material	Thickness (inches)	No. of Tests	Time to Ignition (sec.)	Time to Peak Rate (sec.)	Heat Release Rate, w/cm^2 Peak	1 Min. Av.	5 Min. Av.	Total Heat Released Btu/ft. 5 Min.	10 Min.	15 Min.	20 Min.
Exposure: 2.6 w/cm^2, pilot flame bottom edge											
Lumber: D. Fir	1.5	1	26	70	9.7	4.2	4.0	1050			
Plywood: Sp. D. Fir, untr.	.625	2	13	17	11.5	9.0	7.0	1860	3420		
Sp. D. Fir, tr.	.625	2	14	19	8.0	2.6	2.6	690	1410		
Exposure: 2.6 w/cm^2, pilot flame top edge											
Lumber: D. Fir	.75	2	98	107	7.8	0	3.1	804	1735	2403	
So. Pine	1.5		130	180	10.3	0	3.2	840	2490	3690	4770
Redwood	1.5	1	270	335	8.6	0	0	0	1200	2160	3000
Exposure: 4.0 w/cm^2, no pilot											
Lumber: D. Fir	1.5	3	65	70	10.4	0	5.4	1480	3100	4500	5800
So. Pine	1.5	5	88	100	11.6	0	7.6	2220	4020	6000	6900
Redwood	1.5	4	51	62	10.0	1.7	4.3	1280	2510	3720	4710
Plywood: Marine D. Fir	.75	3	63	80	14.9	0.4	6.1	1650	2870	4200	5760
Sp. D. Fir, untr.	.625	1	66	70	11	3	5	645?			
Sp. D. Fir, tr.	.625	1	20	390	2.9	0.9	1.6	420			
Sp. D. Fir, tr.	.625	2	a/	405	4.4	0.4	1.3	510			
Exposure: 6.0 w/cm^2, no pilot											
Lumber: D. Fir	1.5	3	13	40	12.4	8.9	8.0	2130	3860	5430	7085
Sp. D. Fir, untr.	1.5	2	12	27	11.8	7.3	9.1	2400	4560	6440	8100
Sp. D. Fir, tr.	1.5	2	a/	10	3.1	1.2	1.89	435	1170	1935	2820
So. Pine	1.5	6	17	81	12.6	9.2	10.5	2856	5172	7002	9080
Redwood	1.5	5	12	21	11.1	9.1	7	1850	3340	4830	5950
Plywood: Marine D. Fir	.75	3	8	19	11.9	9.8	8.2	2180	3860	5520	7380
Sp. D. Fir, untr.	.625	2	13	17	11.5	9.0	7.0	1860	3420	4770	
Sp. D. Fir, tr.	.625	2	14	19	8.8	2.6	2.6	735	1500	2520	2880
Exposure: 8.0 w/cm^2, no pilot											
Lumber: D. Fir	1.5	3	4	38	8.0	5.2	6.3	1660	3240	4560	5880
So. Pine	1.5	3	7	70	11.1	7.1	9.3	2470	4640	6390	7210
Redwood	1.5	4	5	34	9.4	6.5	5.4	1840	3450	4770	5970

a/ Smouldering only

TABLE 2: Fuel Consumption in ASTM E119 Fire Endurance Tests for Wood-and Steel-Stud Wall Assemblies

Time period	Average Fuel Consumption Rate, CFM (total used, cubic ft.)					
	Wood-stud Gypsum board One layer 8 January, 1976	"Reference" Walls: Gypsum Board on Steel Studs				
		One layer Gypsum board 11 March, 1976	Two layers Gypsum board 17 March, 1976	Two layers Gypsum board 23 March, 1976	Two layers Gypsum board 5 April, 1976	Two layers Gypsum board 21 October, 1976
0-15 minute	72 (1100)	70 (1050)	74 (1100)	78 (1170)	76 (1140)	77 (1160)
0-30 minute	83 (2500)	72 (2160)	78 (2340)	78 (2340)	80 (2400)	81 (2430)
0-45 minute	81 (3500)	72 (3240)	75 (3380)	76 (3480)	80 (3600)	83 (3735)
0-60 minute	79 (4700)	71 (4260)	73 (4400)	77 (4600)	80 (4800)	82 (4920)
15-60 minute	80	71	73	76	81	85

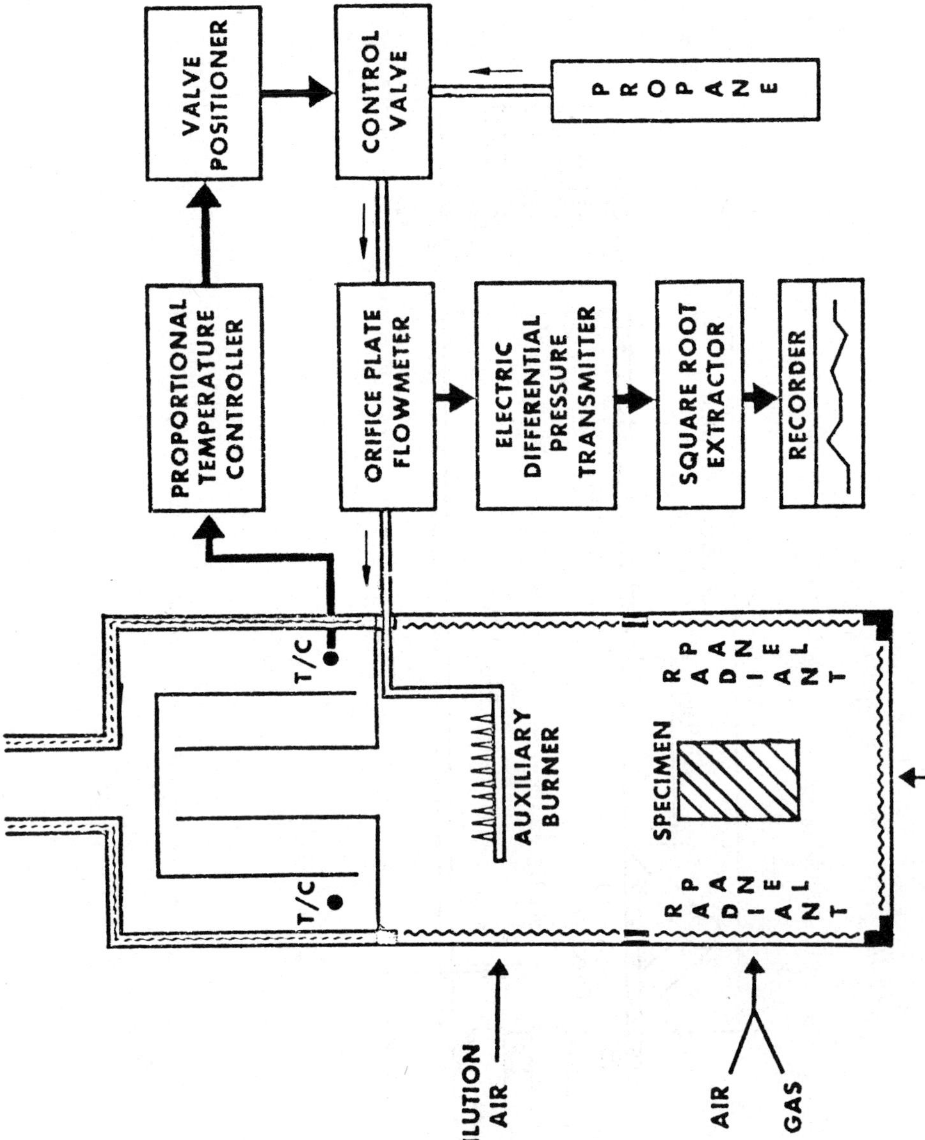

Figure 1. Schematic diagram of NBS HRR calorimeter.

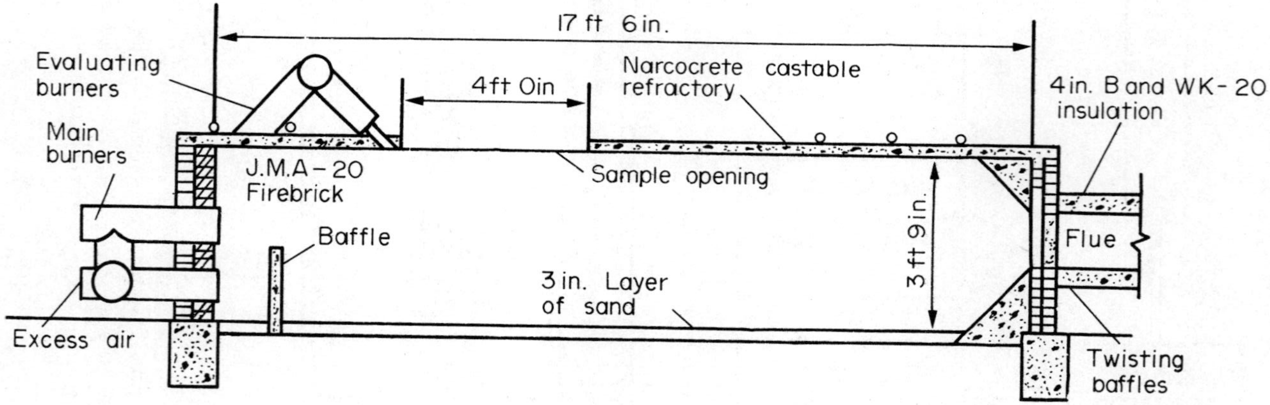

Figure 2. FM construction materials calorimeter.

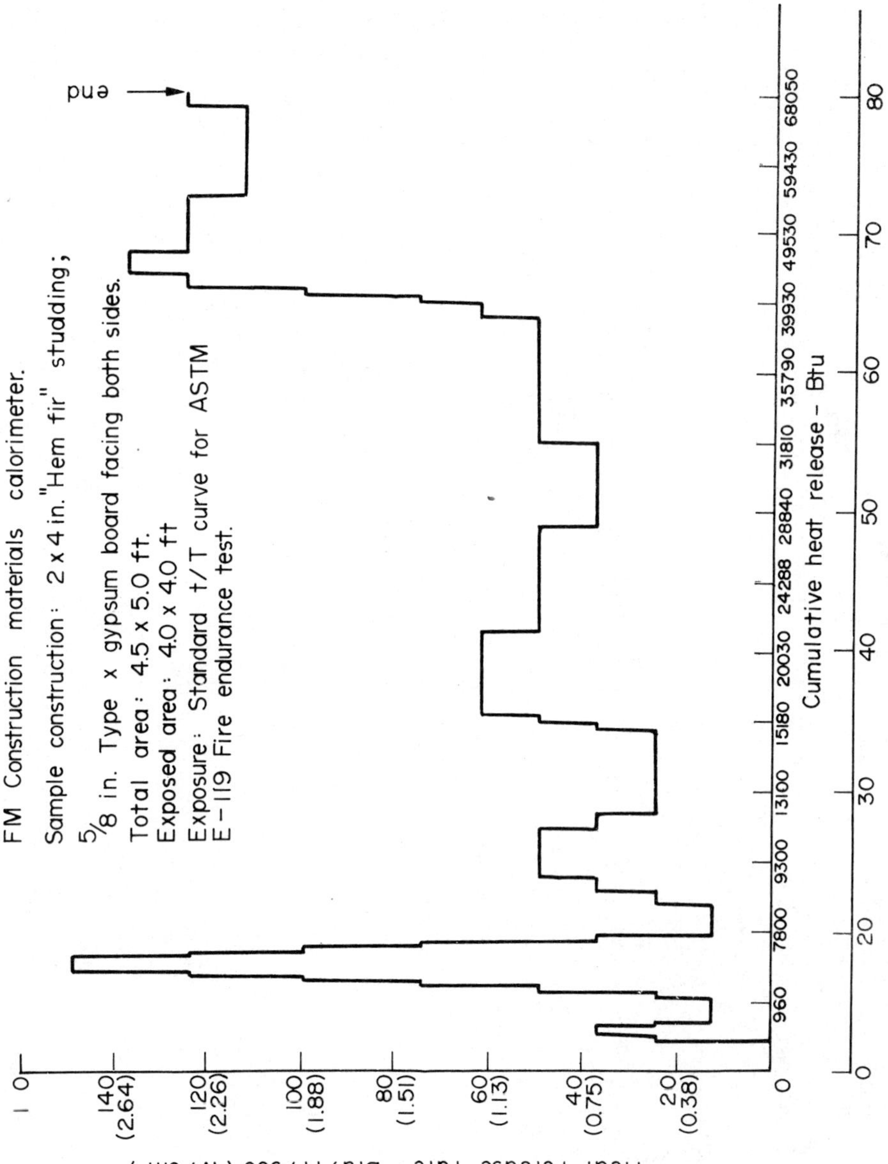

Figure 3. Heat release rate from a model wall assembly.

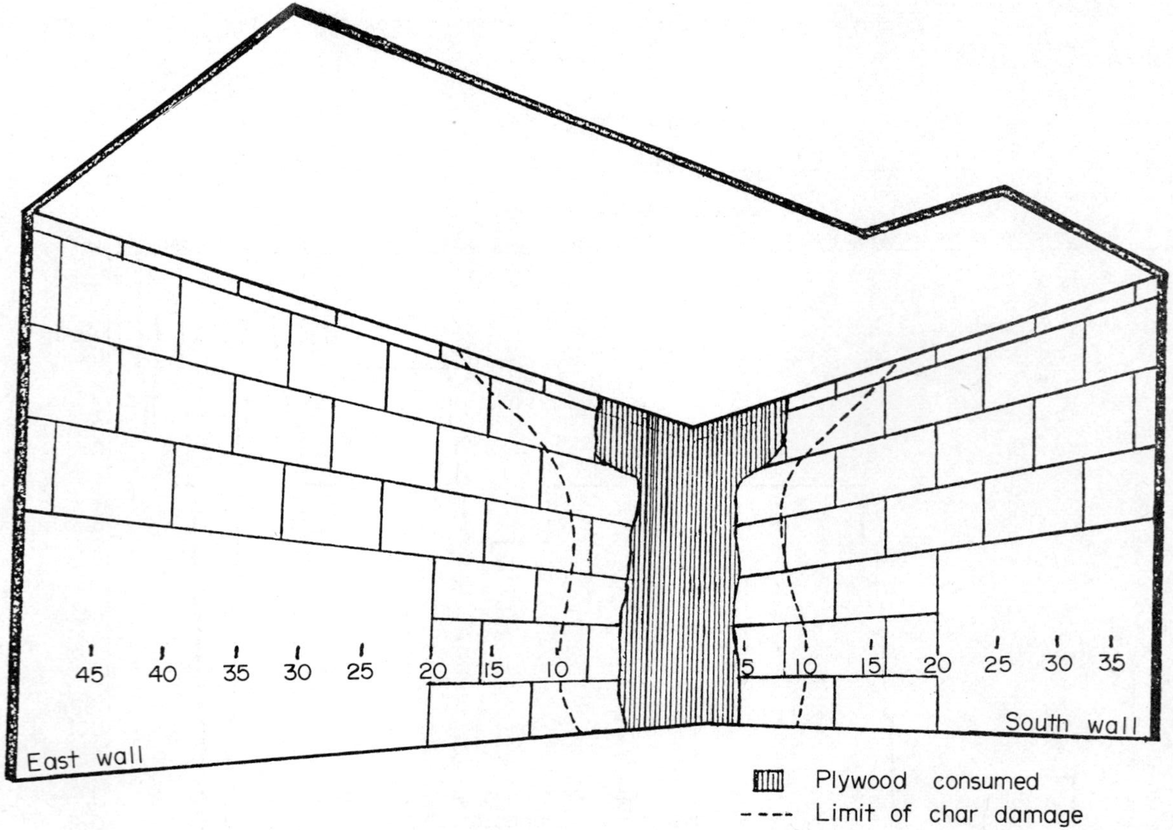

Figure 4. American Wood Preservers Institute comer test. December 19, 1974, ½ inch fire retardant treated CDX plywood.

Main Trends and Technical Problems in the Production of Fire-resistant Wood-based Panels in the USSR

Y.S. Tupitsyn
Departmental Deputy Chief, Ministry of the Lumber Industry of the USSR

V.I. Biryukov
Director of the All-Union Scientific Research Institute of the Woodworking Industry

SUMMARY

The increased use of wood particle and wood-fibre board in industrialized building has led the State fire supervisory bodies of the USSR to introduce more stringent fire-safety requirements for these materials. The use of certain types of wood-based panels which are available on the market is not generally allowed in building, as they do not meet the higher requirements.

A considerable number of methods are known in the USSR for manufacturing fire-resistant panel materials based on comminuted wood. However, they all fall into the following three main categories:

- surface treatment of the panels
- impregnation, and
- introduction of fire retardants into the compressed mass.

Fire retardants used in the manufacture of fire-resistant wood-based panels must satisfy a host of requirements: they must display high fire resistance, low toxicity and chemical stability, and must not cause corrosion of equipment, etc.

In the USSR, coatings containing sodium metasilicate as the main bond (SK-L, SK-G, paints) organochlorine bonds (MKhS, PKhVO-A, KhZN, etc. paints) and coatings based on phosphate bonds - which also include swelling agents, fillers and pigments - are the most widely used. A technique has been developed for producing wood-particle board with vermiculite coatings.

Endowing panels with fire-resistant properties at the production stage represents the most promising development in solving this problem. However, introducing fire retardants into the compressed wood mass frequently reduces the physical and mechanical parameters of the finished product. This is confirmed, for example, in the manufacture of wood-particle board based on urea-formaldehyde resin and a mixture of ammonium phosphate, ammonium sulphate and boric acid, and in the manufacture of "Ramolit"-type wood-fibre board, whose composition includes the wood fibre mass, the water-insoluble binary salts of phosphoric acid and other additives.

The dry production method offers the most favourable possibilities of obtaining fire-resistant panels with satisfactory physical and mechanical characteristics. In the USSR, a technique has been developed for manufacturing wood-fibre panels in which a fire-resistant compound based on orthophosphoric acid, urea and dicyandiamide is introduced into the fibre. Although the finished panels have good physical and mechanical characteristics, their widespread use is prevented by a number of shortcomings.

In view of the high fire-safety requirements to be met by building materials, it is impossible to produce cheap fire-resistant wood-based panels with more or less multi-purpose characteristics. In developing new panels, it would be advisable to start out by bearing in mind their specific function, and try to ensure a set of properties suited to operating conditions in specific building structures.

(The complete text of this paper, in Russian, will be supplied on request by the ECE/FAO Timber Division, Palais des Nations, CH-1211 Geneva 10, Switzerland.)

Fire-resistant Fibreboard

Edmund Urbanik
Wood Technology Institute, Poland

Up to now, the methods used to reduce the combustibility of fibreboard consisted mainly of surface treatments with various fire retardants. These are costly and, frequently, not efficient enough. It was also found that protective treatment cannot be, for various reasons, performed during commercial production processes.

After several years' investigation in the laboratory, a pilot plant on a commercial scale, carried out, at the Wood Technology Institute, a new method of producing insulating fibreboard, characterized by reduced flammability. This new production method eliminated most of the difficulties and drawbacks with regard to fire-protection of ready-made insulating fibreboard.

The reported study was aimed at:

a/ Selection of fire-retardants, determination of their chemical composition, and means of mixing them with the groundwood

b/ Experimental treatment of fibreboard under laboratory, pilot plant and commercial production conditions

c/ Determination of physical and mechanical properties of fibreboard treated with fire-retarding compounds

d/ Evaluation of the effectiveness of fire-retardant treatment

e/ Elaboration of principles governing the production technology in the manufacture of slow-burning fibreboard.

On the basis of experience gained in the laboratory and pilot-plant experiments, a few batches of fire-resistant fibreboard were prepared under commercial scale conditions. It was found that for continuous production on a technical scale equipment additional to existing production lines was necessary. Such equipment should not, of course, interfere with the actual production of ordinary insulating fibreboard.

The essence of the new technology of slow-burning fibreboard production consists of the addition of chemical fire-retardants, in the form of aqueous solutions, to the wet fibreboard mat using a specially developed proportioner and suction device. Negative pressure and amount of metered fire-retardant are controlled, depending on the kind of chemical used and the thickness of fibreboard produced, to obtain a thorough saturation of the mat. The filtrate is not drained to a sewer, but re-circulated in a closed system and used again for the preparation of the next batch of treating solution. The treated fibreboard mat travels then to the sizing section of the line and thence to the drier. After cutting to the desired dimensions, fibreboard is ready to be used as slow-burning insulating board or, after proper preparation, pressed in a hydraulic press to obtain slow-burning hardboard.

Some physical and mechanical properties of the new wood-based product

In Table 1, results are presented of the tests on physical and mechanical properties of slow-burning fibreboard, prepared according to the developed technology under commercial production conditions, as well as untreated fibreboard obtained from current industrial production. It can be seen from this table that hygroscopicity of treated material under normal exploitation conditions, at about 65% relative humidity of the air, does not differ essentially from the hygroscopicity of untreated boards. Water absorption capacity of treated fibreboard during soaking is considerably higher than the capacity of untreated material. This property should not, however, constitute the criterion of quality assessment, because fibreboard is, from the nature of things, not tolerant of free water action and, therefore, is not intended for external use.

Effectiveness of treatment with fire-retardants

Flammability tests with fibreboard were carried out at the Wood Technology Institute with the application of the following methods:

1. Thermal radiation technique (Prüfstrahler)
2. Gravimetric-ignition method developed in the Wood Technology Institute
3. Schlyter method
4. British thermal radiation method according to BS-476.

Insulating fibreboard samples, treated and untreated (controlled), prepared according to technical principles elaborated in the Institute both under laboratory and commercial production conditions, were subjected to testing. Test results demonstrated that treated samples did not ignite even after 15 minute exposure to the action of fire, irrespective of the test method applied. Furthermore, no spread of flame on sample surfaces was observed. On the other hand, control samples ignited, under similar conditions after 6-8 seconds of exposure to the fire source.

On the basis of current criteria, it can be said that insulating fibreboard and hardboard, fire-proofed by the method developed in the Institute and produced under commercial conditions, is a slow-burning material characterized by the absence of surface spread of flame.

TABLE 1. PHYSICAL AND MECHANICAL PROPERTIES OF EXPERIMENTAL FIBREBOARD TREATED WITH FIRE-RETARDANTS

Properties		Unit of measure	Insulating fibreboard		
			From the current production of industrial plant	Experimentally prepared on commercial scale, and treated in the course of production process with fire-retardant based on:	
				phosphoric acid and amine resin	ammonium salt of phosphoric acid
Thickness		mm	12.6	13.3	13.5
Density	mean	kG/m^3	276	312	294
	minimum			266	272
	maximum			352	336
Static bending strength	mean	kG/m^2	18	18.5	15.4
	minimum			16.2	14.0
	maximum			19.0	17.4
Hygroscopicity after 7 days of conditioning in air of relative humidity:	65%	%	9.4	9.0	10.2
	98%		17.2	25.2	37.3
Water absorption capacity in soaking: insulating fibreboard after 2 hours		%	28	160	245
hardboards after 24 hours			45	280	260

New Concepts in the Fire-retardant Treatment of Wood-based Materials

R.W. Watson M.Sc
Technical Director, Hickson's Timber Products Ltd.,
United Kingdom

INTRODUCTION

For many years man has used wood in association with a variety of ancillary chemical treatments, designed to overcome one or more of the inherent deficiencies of that material. In this way he has been able to make the very best possible use of the many admirable qualities of the oldest of all building materials.

More recently, research into methods of wood treatment has intensified dramatically, as a result of increasing economic and environmental pressures. Economic considerations dictate that the effective service lives of wooden components should be maximized and optimum utilization of material be achieved, often by encouraging the use of man made wood based products. Environmentalists demand that the chemicals and treatment systems used for wood processing should be as safe as possible and that the use of products which contribute to increased safety be encouraged.

As a result the use of fire retardant treatments for wood based materials of all kinds is now being actively encouraged, often through the medium of Building Regulations.

The main impediment to the increased use of fire retardant treatments has been the nature of the products themselves. Commercial formulations are normally based on inorganic salts such as phosphates, sulphate, borates and halids which are impregnated into the wood in aqueous solutions and which remain essentially water soluble in service. This means that the products, many of which are hygroscopic by nature, are susceptible to fluctuating moisture conditions and may give rise to problems with overpainting, gluing and corrosion. In addition more recent evidence has indicated that treatments of this nature may also have significant adverse effects on the strength properties of wood, particularly where high temperature post treatment kiln drying has been employed. Direct reference to these problems was first made by Gardner (1965) in the United States and has been repeated more recently by Laidlaw and Brazier (1974).

While conventional inorganic fire retardants applied at suitable loadings, can often provide the degree of fire protection called for by the various National and Local regulations, the ancillary difficulties referred to have restricted their use and often resulted in wood based products being superceded by other materials. It is for this reason that extensive research has been carried out into this problem, as a result of which the product Non-Com Exterior (NCX) was developed, first of all as a treatment for Cedar roofing shingles in the United States.

NEW FIRE RETARDANT

NCX represents a radical change in fire retardant treatments, particularly from the standpoint of resistance to weathering and the problems of hygroscopicity. The material for treatment is impregnated with NCX chemicals in a variety of ways ranging from conventional vacuum pressure methods for lumber, to the impregnating of veneers for plywood or the spraying of chips in the manufacturing of particle board. After the treatment cycle, it is required to polymerize the chemicals by heating. This may be carried out in a dry kiln, veneer drier or chip drier.

After the polymerization is complete, the fire retardant is locked into the wood fibres and this gives the resultant material its exterior or anti-weathering properties. Furthermore, the polymerization also causes the nature of the chemicals to change and thus eliminate entirely the dangers of increased hygroscopicity of the treated material. The treated material can then be finished in the normal manner by painting, staining, varnishing, veneer overlaying, polishing or indeed any other methods presently used in the industry.

FIRE RETARDANT PERFORMANCE

Before fire retardant treated material is submitted for test, careful consideration must be given to the nature of the test method and it is often advisable to carry out a number of "sighting experiments" before applying for the official examination. Experience has shown that the loadings of chemical and concentration gradient in the wood required to pass, for example, the BS 476 Part 7 Spread of Flame test are completely different from those required to fulfil the requirements of the Swedish B4 Hot Box test. The question of concentration gradient is one to which particular attention has been paid in the development of NCX treatment; so much so that gradients can now be carefully built into plywood destined to satisfy a given regulation, by treating the individual veneers to different retentions.

NCX treatments have been approved in the United States resulting from tests carried out by Underwriters' Laboratories Inc and current listings for NCX treatments on lumber and plywood give an FR-S rating for Douglas fir, Southern pine, White fir, Western hemlock, Redwood (Sequoia spp) and Ponderosa pine lumber; western species of plywood, Southern pine and Douglas fir plywood. In addition most of these species have been subjected to an accelerated standard rain test, equivalent to 800 inches of rainfall over a 10 year period and have produced no reduction in fire performance.

In Europe tests are being carried out according to different National Standards requirements. In the UK the BS 476 Part 7 test measures the spread of flame across the sample surface and Class 1 ratings have been achieved for Baltic Redwood, Spruce, Western Red Cedar, Douglas fir plywood and combinations of Birch and Spruce plywood.

Class 0 ratings on both Douglas fir plywood and Western Red Cedar have been achieved in the BS 476 Part 6 test, despite the well-known variability problems associated with the Fire Propagation apparatus.

In France an M1 rating in the CSTB Epiradiateur test (J O Francoise 1011) was obtained on Gaboon plywood.

Swiss tests carried out on Spruce plywood at EMPA gave the best classification for inflammability and smoke intensity.

Tests in Germany to DIN 4102 on Spruce cladding gave pass levels for both surface destruction and flue gas temperature in the Schwerantiflammbarkeit.

Finally tests carried out in accordance with the Danish version of the Swedish Hot Box technique SPA-8 gave a Class 1 rating.

ANCILLARY PROPERTIES

Leach resistance

The original reason for the development of NCX fire retardant was the growing realization of the need for an effective fire retardant which retained its effectiveness for extended periods under conditions where leaching was a hazard.

In order to demonstrate this property of the treatment a series of blocks of Baltic Redwood (*Pinus sylvestris*) measuring 3/4" x 3/4" x 3/4" was treated with two fire retardant solutions. These consisted of solutions of NCX and diammonium phosphate, a chemical much used in conventional fire retardant formulations.

After drying and curing, the blocks were subjected to the weathering procedure described in ASTM D1413 (1961) in which a two week cycle is employed to provide alternate wetting and drying conditions. The washings from each test were compounded and analysed for nitrogen and phosphorus. The leaching results, expressed in terms of percentage of each element lost, are summarized in Table 1.

TABLE 1 Loss of nitrogen and phosphorus from NCX and diammonium phosphate

% N lost from:		% P lost from:	
NCX	DAP	NCX	DAP
9.3	82.6	36.5	78.2

These results indicate the marked superiority of NCX compared with diammonium phosphate treatment.

Hygroscopicity

In many of the situations where one would wish to use fire retardant treated wood, leachability in the true sense of the word is not a problem. The natural hygroscopicity of fire retardant salts themselves however, can present a far greater threat to the successful application of fire retardant treatment, by virtue of the practical problems which can result when the treated wood is exposed to fluctuating moisture conditions.

The inorganic salts which form the basis of most conventional fire retardant treatments, are normally hygroscopic at relative humidities of 80-90 per cent dependent upon the ambient temperature prevailing. A series of tests comparing NCX treatment with conventional salts has already been reported by St Clair (1971) and the results after seven days at 100°F and 85 per cent RH are summarized in Table 2.

TABLE 2 Wood moisture content (per cent) after 7 days' exposure at 100°F and 85% RH

Sample Board	Untreated	Phosphate	Sulphate	NCX
A	16.9	26.7	37.3	16.2
B	16.3	28.9	37.4	15.8
C	17.3	27.2	38.2	15.0
D	16.9	28.8	39.3	15.5
E	16.9	31.4	40.5	15.2
F	16.6	30.8	39.1	15.5
Average %	16.8	28.9	38.6	15.5

These results, in addition to demonstrating conclusively that NCX treatment resembles untreated wood in its hygroscopicity characteristics, also confirm the already established fact that ammonium sulphate is more hygroscopic than the corresponsing phosphate at this relative humidity.

It was also observed during these tests that salt blooming only occurred on the two sets of salt treated samples.

Strength

The strength properties of NCX treated products are essentially unchanged by the treatment and the results of strength tests conducted by independent laboratories are reported in Tables 3 and 4.

TABLE 3 Static bending test results on NCX treated and untreated matched Southern pine lumber

	MOE($x10^6$psi)	MOR($x10^3$psi)	Wml(in-lb/in^3)
Untreated control	1.89 ± 0.40	15.99 ± 2.20	12.81 ± 3.87
NCX treated	2.00 ± 0.37	14.06 ± 2.27	10.15 ± 3.47

In the United Kingdom strength tests have been carried out on NCX treated Scots pine. The results are summarized in Table 4.

TABLE 4 Static bending test results on fire retardant treated and untreated Scots pine

Sap/Heart	Treatment	MOR(N/mm^2)	MOE(N/mm^2)
Sap	Control	97.3	12110
Sap	NCX	120.7	12860
Heart	Control	90.1	9910
Heart	NCX	92.8	8200

Gluing

Soon after the development of NCX treatment, it became evident that this product had great potential in the laminating industry. This interest, primarily on the part of architects, stems from their desire to design using the inherent qualities of wood without sacrifice of fire safety. In many areas, insurance rates tend to discourage the use of untreated glued laminated wood or heavy timber construction, with substantial discounts given to fire retardant treated members or non-combustible materials.

During early testing it was found that the NCX treatment interfered with the curing of commercially available resorcinol adhesives resulting in very low quality bonds. This prompted the development of a specially formulated resorcinol adhesive designated as Koppers RDL-49.

This adhesive when used in accordance with manufacturer's instructions results in high quality face bonds. Since resorcinol adhesives are considered wet use adhesives by the American Institute of Timber Construction (AITC), quality of the bonds is determined through block shear and cyclic delamination tests. Shear strength values of 1,400 to 1,500 psi with 85 to 90 per cent wood failure and cyclic delamination values (after two cycles) below four per cent are readily achievable. These results, tested in accordance with AITC standards and United States Commercial Standard for Structural Glued Laminated Timber, CS253-63 (Table 5) have been consistently reproduced in laminating plants.

End jointing of lumber is accomplished through the use of radio frequency (RF) cured melamine adhesive applied to finger-joints. A very minor modification in the cycle time allows for a complete cure of the adhesive. These joints have been tested in bending, tension, and cyclic delamination and meet the requirements of the AITC and United States Commercial Standards.

Bending and delamination values equal those for untreated wood. The tensile strength, however, is reduced by 20 to 25 per cent. This reduction is of the same order of magnitude as the reduction in tensile strength in clear wood samples that have been treated with NCX fire retardant.

TABLE 5 Typical results comparing NCX and untreated Southern pine

	Untreated	NCX Treated	AITC Requirement
Face bonding (RDL-49 adhesive)			
Block shear			
psi	1590	1475	1180
per cent wood failure	85	85	70
Cyclic delamination (2 cycles) per cent delamination	0.5	3.2	10.0
End bonding (RF cured melamine adhesive)			
Bending			
psi	15100	14200	8190
per cent wood failure	95	90	70
Tension			
psi	9550	7580	8190
per cent wood failure	77	82	-
Cyclic delamination	1	3.5	10.0

More recently attention has been focussed on the gluing up of NCX treated veneers. Experimental work, now substantiated by commercial scale experience, has revealed no difficulties in bonding with conventional phenol and melamine formaldehyde adhesives to the exacting standards of the Finnish Plywood Development Association, using orthodox glue spreading techniques.

Preservative qualities

The resistance of NCX treated wood to attack by wood destroying fungi has been the subject of detailed laboratory and field investigations.

When tested in accordance with CEN draft document PR/U7, which closely resembles BS 838 (1961), the effectiveness of the treatment is shown against fungal attack. A range of NCX retentions was evaluated and the results are shown in Table 6.

TABLE 6 Toxic limits for NCX expressed as retention of preservative in wood (kg/m3)

Fungus	Retention (kg/m3)
Coniophora cerebella	$<$ 35.2
Poria monticola	$<$ 36.2
Lenzites trabea	$<$ 36.0
Chaetomium globosum	30.7 - 47.6

The results for all four fungi indicate that good resistance to fungal attack can be achieved well below the normal commercial chemical retentions employed.

TABLE 7 Performance of NCX and untreated 3/4 inch stakes in test at Charleston, South Carolina, USA

Preservative	When Installed	Species	Index of condition (years)						
			1-2	2-3	3-4	4-5	5-6	6-7	7-8
NCX	1964	SYP(1)		85	84	63	60	44	38
Untreated	1964	SYP(1)	27	13	0				
NCX	1967	WRC (sap) (2)	100	100	100	100			
Untreated	1967	WRC (sap) (2)	100	68	30	11			
NCX	1967	WRC (heart) (2)	100	100	100	100			
Untreated	1967	WRC (heart) (2)	85	0					
NCX	1967	Redwood (3)	100	100	100	100			
Untreated	1967	Redwood (3)	100	78	61	35			
NCX	1967	Douglas Fir	96	80	77	75			
Untreated	1967	Douglas Fir	67	22	8	0			

Key - (1) SYP - Southern Yellow Pine
(2) WRC - Western Red Cedar
(3) Redwood (Sequoia)

Field test data in accordance with ASTM Standard D-1758 has been obtained and Table 7 shows the "Index of Condition" of NCX treated stakes compared with untreated stakes after different number of years of exposure. It is quite clear that a considerable degree of protection is provided by the NCX treatment.

Corrosion

The effect of NCX treated wood on various metal fixings has been investigated. The fittings included steel, copper, sheradized, galvanized, brass and aluminium nails, screws or hinges. Measurements of change of weight of the fittings were made in addition to visual inspections over a period of 18 months in outdoor, out of ground contact exposure. No adverse effects on any metal were noticed.

APPLICATION

The properties of NCX described above illustrate that the benefits of NCX fire retardant treatments arise because of the ancillary properties of the treated wood. Because of the leach resistance, non-hygroscopicity and ability of the treated wood to be laminated, the use of fire retardants can now be extended. In the USA the following commodities have been effectively treated with NCX:-

(a) Cedar shakes and shingles

(b) Cooling tower lumber

(c) Lumber and plywood for non-bearing exterior walls

(d) Lumber and plywood for unexposed applications in areas having high relative humidity

(e) Railway bridge sleepers and timbers

(f) Subway and elevated railroad sleepers

(g) Waiting platforms.

In Europe the principal outlets up to the present time have been the incorporation of NCX into decorative plywoods either at the veneer stage before laminating or treatment of the made up board. In addition treatments of exterior house cladding have been carried out and investigations are also being made into the treatments of roof truss material for housing and industrial or agricultural pole buildings. In the latter cases the non-corrosive nature of the treated wood is essential when truss plate construction methods are used.

The scope for such a treatment is exciting and it seems clear that many applications exist which have yet to be considered by the organization marketing this process. What is certain is that NCX represents a significant step forward in environmental control in the contribution which it can make by providing a stable long lasting protection against the ravages of fire.

REFERENCES

1. Gardner R E (1965) The ancillary properties of fire retardant treated wood. Forest Prod. J. 15 (9) 365.

2. Laidlaw R A and Brazier J D (1974) The implications of using inorganic salt fire retardant treatments with timber.

3. St. Clair W E (1971) Utilization of fire retardant decay resistant treatments. Proc. AWPA 67 121-125.

Means to Evaluate Panel Materials in Fire and the Role of Fire Retardants

Birgit A-L Östman and Ernst L Back
Swedish Forest Products Research Laboratory, Sweden

INTRODUCTION

For the evaluation of panel materials in fire, it may be useful to distinguish two aspects. One is a better understanding of the behaviour of panel materials in different fire situations. The other is to find the most equitable classification of panel materials according to their fire behaviour. Both aspects reflect a need for long-term development. Most urgent, however, is the need for an international methodology to classify panels with respect to fire, replacing the many different national methods today. Such work is in progress within the ISO. It is important that different types of existing knowledge, basic as well as applied, should influence the work. The goal is to make these methods widely acceptable and most versatile.

A full understanding of the behaviour of panel materials in different fire situations requires research work, especially on the initial part of the fire, which also includes other interior materials such as textiles and their interaction. This research involves the formation of models based on defined parameters, well known ones e.g. ignitability, rate of heat release, flame spread, smoke or new ones scientifically better defined. Such models should make it possible to predict the development of a fire under different conditions and the relative influence of the defined parameters. They should or might also make it possible to predict the interaction of different materials.

HOW CAN PANEL MATERIALS IN FIRE BE EVALUATED TODAY?

For the best characterization, existing knowledge has to be used. Parameters have to be chosen e.g. ignitability, heat release rate, smoke release rate, all known to have some relation to full scale fires. It is only possible to characterize the fire behaviour of a material by a number of parameters in the same way as a number of parameters are needed to characterize the strength properties of a construction material. It would be useful if several parameters could be evaluated simultaneously in a single test, but they should be reported separately. The relative importance of different parameters is dependent on the real fire situation.

These parameters have to be determined for at least some fire conditions, that is for several levels of radiation intensity, ventilation and geometry. Their correlation to some normal fire situations should be well established. Such testing in different fire conditions will give a more comprehensive and versatile picture of different materials intended for a certain use, e.g. for wood as compared to plastics and to non-combustible materials. It will also make the result more applicable to different fire situations. The Dutch method of time to flashover (NEN 3883) is one of the few existing methods which tests under different radiation conditions. The choice of the fire conditions may be somewhat arbitrary, but it is important that the behaviour of a wide range of materials be well illustrated by the results. As more knowledge is gained, a more diversified picture may be applied.

Some rather obvious criteria for all test methods, such as reproducability, relative simplicity etc, will not be discussed here. But one desirable feature of all these methods may be mentioned. They should express the parameter or the test result according to a continuous scale and not just as a division in classes. Such classes can, of course, be defined later.

The evaluation of panel materials in fire, briefly outlined above, is not very simple and can not be expected to be so since it must reflect a number of complex fire situations with different probabilities. It is unlikely that most nations will enact more fire regulations. But one of the benefits from international test methods is that the test results from many countries can be used and compared, thus saving some effort. Naturally, these international methods will also improve the trade.

WHAT IS IMPORTANT TO REGULATE BY BUILDING CODES?

What is, then, important to regulate as far as panel materials in buildings are concerned? It must be to allow enough time for all human beings to escape. This means that a rapid increase in temperature and an evolution of dense smoke and toxic gases must be delayed or avoided for a sufficient time period. For wood materials, it is probable that the evolution of carbon monoxide is the most serious problem at low radiation intensities. The temperature increase and the smoke evolution are relatively low for wood-based materials, and they can be delayed or reduced either by an increased density, see Fig. 1, or by the addition of flame retardants (Figs. 7 and 8). But the evolution of carbon monoxide is in many cases increased by the addition of fire retardants (Tsuchiya, 1970). As a simple example, the following calculation can be made. For a room, about 3 by 4 by 2 meters, a complete degradation of about 10 kg of wood is enough to give an air content of about 3000 to 4000 ppm carbon monoxide in the air, if the ventilation is neglected. This is lethal to human beings in less than one hour (Sumi, 1971). For a rapid fire development it is probable that the increase of air temperature is more critical, which may be evaluated as e.g. heat release rate.

On the other hand, the rules for panel materials should not be too complicated, since it is known that for most fire situations the interior decoration, furniture and equipment is most important for the initial fire situation in a room. As to escape routes for people, the most stringent rules and regulations should exist.

THE RELATION BETWEEN SOME EXISTING FIRE TEST METHODS

Some of the existing fire test methods for building materials measure different parameters. But even when they measure the same or a closely related parameter, the fire test conditions such as radiation intensity, ventilation, sample size and geometry are often different. In Table 1 some rather similar "box-type" methods are summarized. The supplied effect is calculated as a mean over the total sample area facing the radiation source, although this effect is not evenly distributed over the sample area. Some parts of the sample may receive much higher radiation intensity than the average, especially in some of the methods. Part of the effect is also absorbed by the other surfaces in the box than those of test samples. But it is interesting to note that the calculated mean radiation intensity is of similar order of magnitude, though local variations probably differ more.

TABLE 1 Fire test conditions for some "box-methods"

	Mean effect per sample area kW/m^2	Air supply per box volume min^{-1}	Number and size of samples cm	Measured properties and their evaluation
Scandinavia				
"Hot box" Br-4	18	14	4 pieces 23 x 23	variation of gas temperature and smoke with time
United Kingdom				
Fire Propagation, BS 476, Part 6	10-44	not controlled	1 piece 23 x 23	time dependent index calculated from gas temperature
Federal Republic of Germany				
Brandschacht, DIN 4102	32	160	4 pieces 19 x 100	amount of undamaged test material
Netherlands				
Contribution to flashover and ignition NEN 3883	8-24	not controlled	2 pieces 29 x 29	necessary radiation to produce flashover after 15 min.
Belgium				
"Reaction to fire"	33	not controlled minimum 1	1 piece 30 x 40	total index 1/
France				
Essais par rayonnement	30	not measured	1 piece 30 x 40	4 indices 2/
USA and Canada				
25 foot tunnel ASTM E-84	23	10	1 piece 50 x 760	3 indices based on flame spread, fuel contribution smoke

1/ based on calorific potential, inflammability, flame propagation, flashover sensitivity, rate of heat release and smoke properties

2/ based on ignition time, average and maximum length of flames and average temperature.

The table also illustrates a considerable variation in ventilation conditions between these methods. For some of them the ventilation is not controlled, which means that materials which burn more rapidly also get more air supply by draught. This will probably affect the difference between rapidly and slowly burning materials when tested with controlled or uncontrolled ventilation. A rapid ventilation will also have a tendency to cool the fire.

A third difference between test methods in Table 1 is the number and size of samples. This will influence e.g. the spread of flame from one room to another, the height of flames etc. The absolute size of the samples is probably not as important as a reasonable relationship between the sample size and the air supply or ventilation. Another related difference is the geometrical arrangement of samples and of radiators and burners.

It is well known that, for example, a low radiation intensity is more critical for one type of material than for another. A variation in ventilation conditions, geometry and so on will probably have a similar effect. It is thus not surprising that due to the different fire conditions these test methods may classify building materials in different orders. This is not very alarming as the different test conditions, at best, may simulate different fire situations; moreover, different fire situations have to be considered. But for a standard classification method those conditions should be selected which correlate to the most frequent fire situation or situations.

The correlation to full scale fire behaviour is for most of these tests insufficiently known and stated. This is a drawback in international standardization work.

In the last column in Table 1 the measured properties are listed. These properties are in many cases the same or closely related for the "box tests", and according to some of them an index is calculated. Several of the tests measure more than one property simultaneously, which is an advantage.

THE TOTAL COMBUSTION ENERGY

During most fires, only a minor part of the total combustion energy of the organic materials present is released. This is specially true for wood materials since they char. The total heat of combustion is therefore not a suitable measure of the fire behaviour of a material. Even in a fully developed fire, the total heat is seldom released.

This can easily be examplified. The addition of a fire retardant to a wood-based material may delay the ignition and give a slow heat release. But the material is still completely combustible as for example in a calorimeter with high oxygen pressure (ISO method R 1716), as shown in Fig. 2. The addition of a fire retardant will of course give less combustion energy per total weight of treated material, but not per weight of wood material. This is true for most inorganic fire retardants. Some organic retardants may give an even higher combustion energy during this extreme combustion, (Yeh, 1971). But even the combustion in a bomb calorimeter requires more time when fire retardants are present, especially at lower oxygen pressures, see Fig. 3.

One rather curious exception from these findings is wood-wool building slabs. They do not burn at all, even in a bomb calorimeter and even when well powdered. Wood-wool building slabs contain large amounts of inorganic materials such as, cement, magnesite and gypsum, but about 40 to 45% is wood material.

PHYSICAL MEANS OF ACTION FOR FIRE RETARDANTS

The mechanism responsible for fire retardancy effect is not known in detail for most retardants. But it is well known that many of them lower the degradation temperature of the organic material, increase its char yield, decrease its production of combustible gases and change the composition of the volatile products, thereby reducing their total heat of combustion. These effects are often described as chemical action, but at least for some types of fire retardants the physical action is also important. One approach of this type will be described in some detail (Sellman, 1977).

The most important physical action is to prolong the time to ignition through heat absorption by the retardant, through evolution of inert gases from the retardant and through formation of an insulating surface layer. To be able to quantify these effects, it is assumed that a medium density wood fibre building board is treated with 2 kg of flame retardant per m^2. The flame retardants included are borates, boric acid, phosphates and silicates. The board is assumed to be irradiated with an intensity of 15 kW/m^2. Under these conditions an untreated board is ignited after 6 to 7 minutes. The time to ignition is prolonged by 1 to 5 minutes through heat absorption of the different fire retardants as calculated from published thermodynamic data for these flame retardants. The amount of inert gases, mainly water, emitted is up to 2.6 m^3 per m^2 board. The main results are summarized in Table 2. The effectiveness of inert gases in inhibiting ignition is of course dependent on the ventilation around the fire. The formation on an insulating surface layer is more difficult to quantify.

The data below are calculated for a medium density fibre building board treated with 2 kg of flame retardant per m^2. The radiation heat is 15 kW/m^2 and the board ignites at 380°C. The time to ignition for an untreated board is 6 to 7 minutes.

TABLE 2

Flame retardants	Prolongation of time to ignition through heat absorption minutes	%	Volume of inert gases (at 380°C) emitted per m^2 board up to 380°C m^3
Diammonium phosphate $(NH_4)_2HPO_4$ (s)	5	75	1.9
Monoammonium phosphate $NH_4H_2PO_4$ (s)	3.5	55	1.2
Sodium metasilicate $Na_2SiO_3.5H_2O$ (s)	3	45	2.0
Sodium disilicate $Na_2Si_2O_5$ (gl)	1	15	0
Boric acid, H_3BO_3	4.5	70	2.6
Borax, $Na_2B_4O_7.10H_2O$	3.5	55	2.5

PROBLEMS INVOLVED IN THE USE OF FIRE RETARDANTS

Even if, for instance, the time to ignition can be prolonged considerably by the use of fire retardants, there are practical and technical problems involved in their application. Many fire retardants affect the strength properties and the appearance of wood-based panels. If they are water soluble, as many of them are, the moisture resistance of boards is decreased and there is a risk for salt migration to the surface or into and through the board. Water soluble retardants are also difficult to apply in wet processes, if the process water system is not completely closed. Other problems, not so frequently mentioned, are toxicity and corrosion hazards during manufacture as well as during use and during final destruction of the boards. Halogen compounds are specially hazardous, as they may cause corrosion in all three cases. Many of them, especially organic halogen compounds, are highly toxic or have highly toxic degradation products. Halogen based fire retardants also have a tendency to increase the smoke production during fire. Because of all these risks they should be avoided.

FIRE RETARDANT WET PROCESS FIBRE BUILDING BOARD

Fire retardants with at least partial physical action, are suitable for panel materials. This has been demonstrated in work commissioned by Scandinavian mills carried out at the Swedish Forest Products Research Laboratory. The objective was to find technical and economical methods to reduce the fire degradation of wet process fibre building boards, mainly insulating board and medium density board.

A concentration of the fire retardant to the surfaces, see Fig. 4, is most economical. For a wet processed board a concentration to one surface is only necessary and useful if the retardants are added in an open system for processed water. Any after-treatment followed by an additional drying is usually too expensive. The fire retardancy is in Scandinavia usually evaluated by the fire box method (Scandinavian fire test method nr 4) see Fig. 5, which is a relatively well defined and reproducable method with at least some correlation to full scale fire situations.

Figure 6 shows how the distribution of a constant amount of fire retardant (about 1.4 kg anhydrous borax per m^2 of a 10 mm thick medium density fibre building board) affects the temperature-time curve in the box test. With the same amount of fire retardant very different classifications are achieved depending on its distribution over the board thickness. But this type of water soluble fire retardant is less suitable for a permanent treatment.

In Fig. 7 two types of treatment for insulating fibre building board are presented. One is the incorporation of insoluble borates in the top coat of fibre stock. Scandinavian class 2 is achieved in this way. This board is class 1 according to BS 476 : Part 7, and has 4 to 5 minutes to flash-over according to the so called Schlyter method (Scandinavian fire test method nr 2) compared to 1 to 2 minutes for an untreated insulating board. The physical properties of the board are not changed and its appearance is very similar to normal boards. The other treatment in Fig. 7 is a phosphorous flame retardant resin applied on the wet lap before entering the drier. During fire the resin will behave as an intumescent coating and yield Scandinavian class 1. This board is class 1 according to BS 476 : Part 7 and close to class 0 according to BS 476 : Part 6. According to the Schlyter-test flash-over has not occurred after 15 minutes.

In this context it should be mentioned that a new limiting curve according to the hot box method might be introduced in Scandinavia in the near future. Various types of untreated wood based panel materials except insulating board will pass this new limiting curve nr 3.

There are further limitations in the choice of fire retardants for medium density fibre building boards, since they have to stand a pressdrying at 200 to 250°C.

Figure 8 illustrates a one-side treatment during manufacture of such boards. Scandinavian class 1 can be achieved, which in this case corresponds to class 0 according to BS 476 : Part 6.

The costs for these treatments are high in relation to these relatively cheap products. For some of the best treatments the cost may increase by up to 40 to 50%.

A considerable improvement of the fire properties of wood based panel materials can thus be achieved. This will probably affect the fire safety in spaces like corridors and escape routes. But in many spaces the main fire load is made up by more easily ignitable furniture and textile materials, which are far more important for the development of the initial fire.

LITERATURE

1. Sumi, K. and Tsuchiya, Y.:
Toxic gases and vapours produced at fires.
National Research Council of Canada, Div. of Building Research, CRD 144 (1971)

2. Tsuchiya, Y. and Sumi, K.:
Thermal decomposition products of cellulose
J. Appl. Polym. Sci. 14 (1970) 2003-2013.

3. Yeh, K. and Barker, R.H.:
Pyrolysis and combustion of cellulose IV,
Text. Res. J. 41 (1971) 932-938.

4. Sellman, L-G, Östman, B.A-L, and Back. E.L.:
Methods of calculating the physical action of flame retardants.
Fire and Materials (1977). In print.

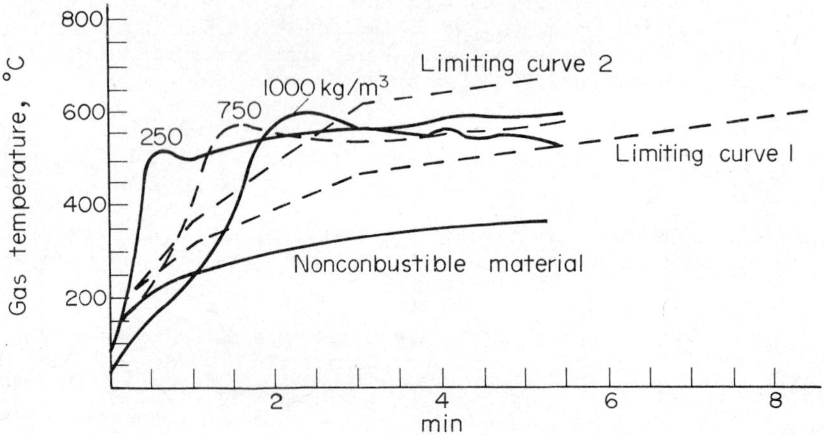

Fig. 1 Effect of density on the fire behaviour of untreated wood-based panels according to the Scandinavian fire test method nr 4 (see Fig. 5). In this method the smoke gas temperature is recorded versus time.

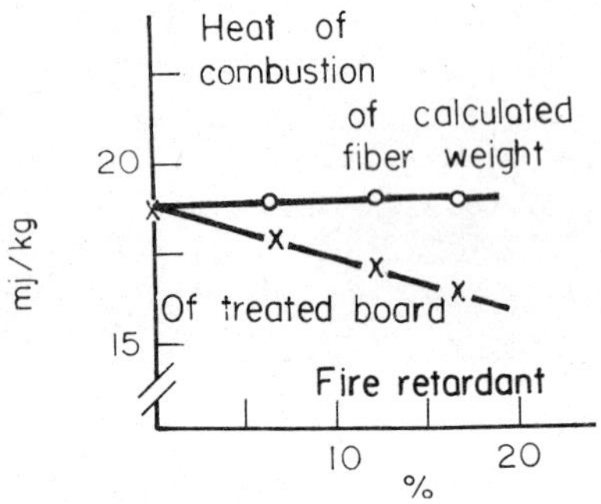

Fig. 2 Total heat of combustion in a bomb calorimeter for medium density board, treated with different amounts of fire retardants, mainly boron compounds.

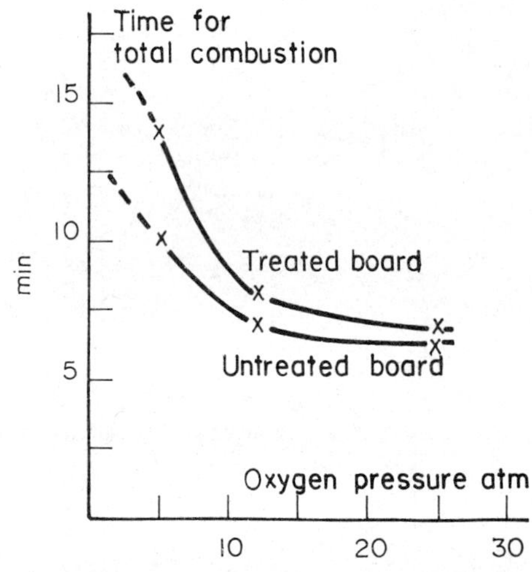

Fig. 3 Time for total combustion in a **bomb** calorimeter for untreated and treated medium density board at different oxygen pressures.

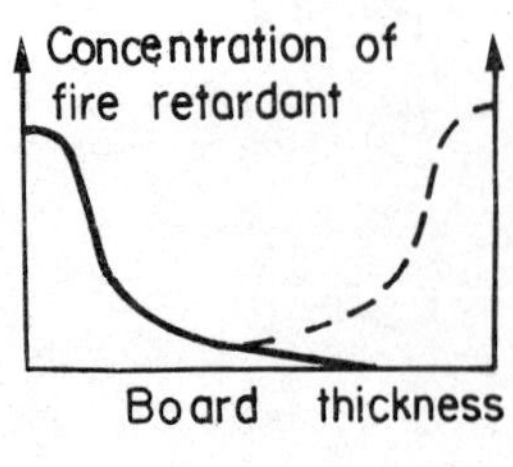

Fig. 4 Economical distribution of fire retardants over the board thickness. The retardant may be concentrated on one or both sides.

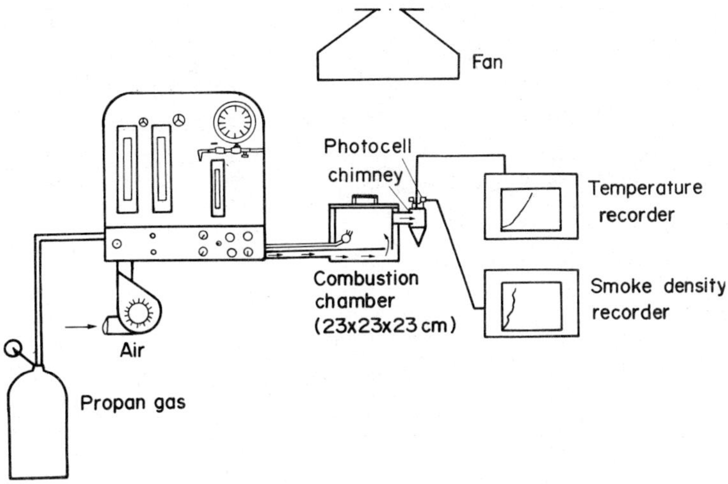

Fig.5 Equipment for the Scandinavian fire test method nr 4, the box test.

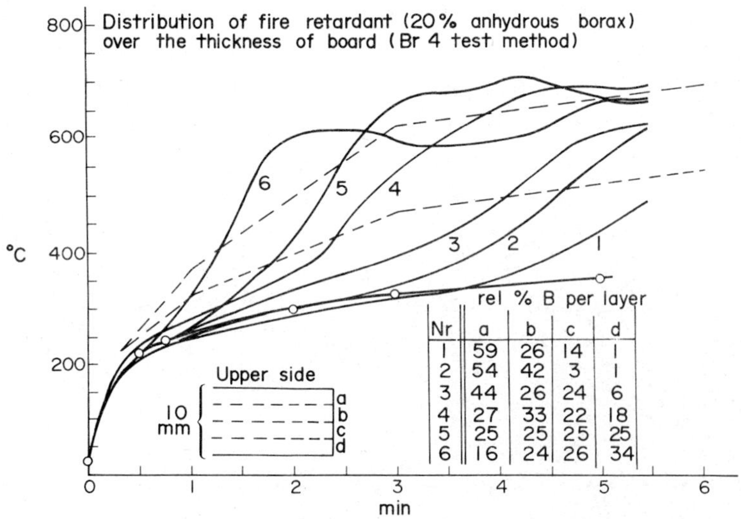

Fig. 6 Effect of different distribution of a given amount of fire retardant over the board thickness. The fire retardancy is evaluated as the temperature versus time according to the Scandinavian fire box method.

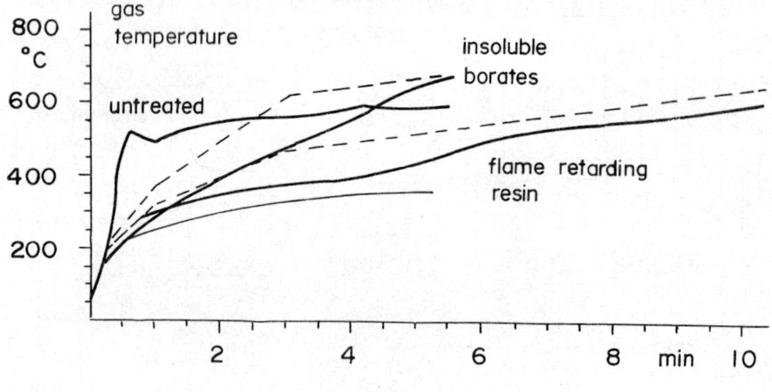

Fig. 7 Insulating board with effective fire retardant treatments on-line according to the Scandinavian fire box method.

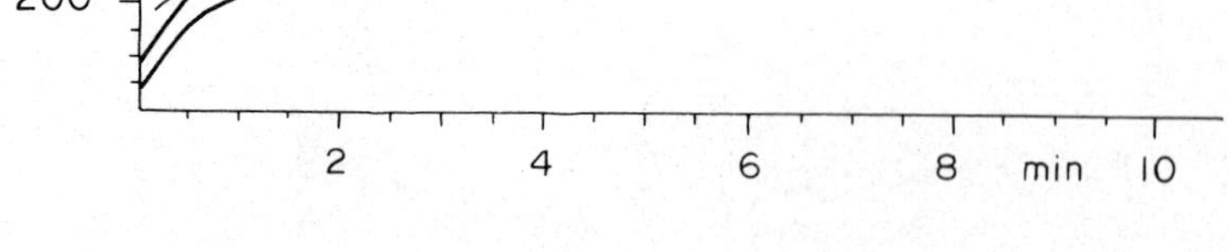

Fig. 8 Medium density board with effective fire retardant treatment on-line according to the Scandinavian fire box method.

Evaluation des Produits Dérivés du Bois Utilisés pour les Revêtements, les Habillages et Accessoires: Problèmes de Toxicité de la Fumée

M. Heusse
Directeur du Développement des "Produits et Procédés nouveaux" des Etablissements Rougier et Fils (France)

Président de la Commission Technique et Feu du Syndicat Général des Fabricants de Panneaux à base de Bois

Le Bois, matériau naturel et ancestral, n'a jamais eu besoin de se justifier aux yeux des hommes, pour être et demeurer le matériau essentiel de son cadre de vie.

Envié, recherché, ou oublié, selon les tendances de l'architecture ou le style du mobilier contemporain, il ne peut laisser indifférent; partout sa présence s'impose, discrète, mais combien complémentaire et nécessaire au monde du minéral, qu'il réchauffe et embellit.

Matériau de structure, aux performances spectaculaires, le bois n'en est pas moins devenu également, ou plus exactement, redevenu un matériau privilégié d'habillage. Sur le plan thermique, l'effet de rayonnement d'une paroi froide, si désagréable, n'était-il pas à l'origine, de ces stalles, lambris ou revêtements en bois massif, habillant les murs de nos anciennes demeures, châteaux ou églises. Dégarnis depuis, ces murs ont retrouvé la pureté du dessin original d'architecture, mais aussi la froideur des monuments historiques rapidement visités et non habités.

Aujourd'hui, grâce aux progrès constants de l'industrie et des chercheurs, le béton a remplacé la pierre. De même façon, les matériaux dérivés du bois, ou les techniques d'usinage en grande série, ont mis le bois à la portée de tous.

Matériau vivant, celui-ci reste nécessaire à l'environnement immédiat de l'homme. Il est par nature source d'intimité, de chaleur, de sécurité.

Il est heureux que nos Pouvoirs Publics aient perçu ce besoin profond, mais il est heureux également que l'industrie, par ses recherches, puisse rendre le bois ou ses dérivés, plus sécurisant encore sous l'aspect incendie dans le domaine des parois verticales auquel nous nous limiterons.

Nous examinerons dans le cadre de cet exposé :

1. La sécurité en matière d'incendie en France
   1.1 Les critères de classement
   1.2 La réglementation française
2. Les panneaux dérivés du bois face à cette réglementation
3. Les domaines d'application et leurs exigences propres

1. LA SECURITE EN MATIERE D'INCENDIE EN FRANCE

1.1 Critères de classement

Les critères de classement requis en façade portent essentiellement sur

Le Pouvoir calorifique en Kcal/kg

Les panneaux dérivés du bois désignés ci-dessus sont assimilés au pouvoir calorifique du bois, soit 4.000 kg/cal. En réalité, les essais ont montré que pour certains types, panneaux de particules ignifués M1 ou M2, il est sensiblement inférieur.

La Réaction au feu : La réaction au feu qui a pour but "d'apprécier l'aliment qui peut être apporté au feu et au développement de l'incendie" (Art. 2 du décret 57-1161 du 17 octobre 1957) par un matériau est déterminée d'après les tests effectués par l'un des laboratoires agréés par le Ministère de l'Intérieur :

- Laboratoire d'essais du Centre Scientifique et Technique du Bâtiment (C.S.T.B.) (Laboratoire pilote)
- Laboratoire Central de la Préfecture de Police
- Laboratoire National d'Essais du Conservatoire des Arts et Métiers
- Laboratoire d'essais du Centre de recherches du Bouchet (Article 1er de l'arrêté du 5 février 1959, modifié le 16 mai 1975).

La combustibilité des matériaux est déterminée à l'aide d'une bombe calorifique. Est considéré M.O. (Incombustible), un matériau dont le pouvoir calorifique supérieur est inférieur ou égal à 600 calories-grammes et à condition qu'il réponde en plus aux spécifications des matériaux M.1 (Art. 67 de l'arrêté du 4 juin 1973).

On distingue deux grandes catégories de matériaux :

1) les matériaux incombustibles classés :
   M.0

2) les matériaux combustibles classés : M.1 (non inflammable)
   M.2 (difficilement inflammable)
   M.3 (moyennement inflammable)
   M.4 (facilement inflammable)
   M.5 (très facilement inflammable)

Les panneaux dérivés du bois font partie des matériaux rigides et à ce titre, sont assujettis à l'essai par rayonnement (essai à l'épiradiateur). Cette méthode d'essai est définie par l'arrêté du 4 juin 1973.

La résistance au feu

La résistance au feu est le temps pendant lequel les éléments de construction peuvent jouer le rôle qui leur est dévolu malgré l'action d'un incendie. (Art. 2 du décret 57-1161 du 17 octobre 1957). Cette résistance est déterminée d'après les essais effectués dans l'un des laboratoires agréés par le Ministère de l'Intérieur.

- Laboratoire du Centre Scientifique et technique du Bâtiment (C.S.T.B.) (Laboratoire pilote)

  Laboratoire d'essais du Centre Technique Industriel de la Construction Métallique (C.T.I.C.M.) (Article 2 de l'arrêté du 5 février 1959 modifié le 24 avril 1972).

Les classements suivants sont établis :

Stable au feu, selon la durée de résistance mécanique au cours de l'essai

Pare-flammes, selon le temps pendant lequl il répond également aux critères d'étanchéité aux flammes et l'absence de gaz inflammables

Coupe-feu, selon le temps pendant lequel, en plus des critères déjà mentionnés, l'élément joue le rôle d'isolant thermique; la température de la face non exposée ne doit pas dépasser 180°C en un point, ou 140°C moyenne des différentes températures relevées, la température ambiante étant toujours déduite des températures relevées.

Un élément de construction peut-être classé "stable au feu", "pare-flammes", ou "coupe-feu", de degré 1/4h, 1h, 1h1/2, 2h, 3h, 4h, ou 6h, selon le temps pendant lequel il a satisfait aux essais.

Comportement au feu des façades vitrées

Un arrêté du 10 septembre 1970 (J.O. du 19 septembre 1970) définit les essais L.E.P.I.R.2 (Local expérimental pour un incendie réel à deux niveaux) et la classification qui en résulte : indice C.

L'indice C est la distance minimum entre les parties non détruites au cours de l'essai situées de part et d'autre du plancher.

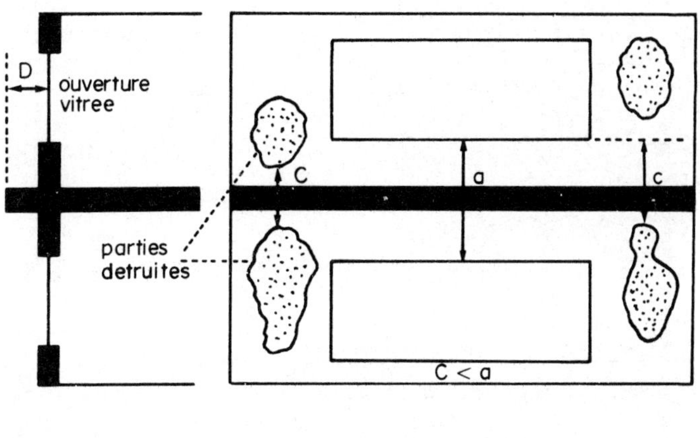

D représente la distance horizontale entre le plan des vitres et le nu de la plus grande saillie de l'obstacle résistant au feu qui sépare les panneaux situés de part et d'autre du plancher.

Matériaux et produits de synthèse

L'arrêté du 4 novembre 1975, complété par celui du ler décembre 1976 et applicable depuis le 10 janvier 1977, réglemente l'emploi des matériaux et produits de synthèse tels que matières plastiques, fibres et textiles synthétiques, élastomères, peintures et vernis colles dont la composition comporte de l'azote et du chlore pouvant être libérés sous forme d'acide cyanhydrique ou d'acide chlorhydrique.

En sont exclus :

- les matériaux classés MO et M1
- les matériaux constitués par un mélange de matériaux naturels et synthétiques (tels que les panneaux dérivés du bois) si la teneur en matière synthétique n'excède pas 25 %.

Le fabricant est alors tenu de garantir cette teneur maximale de 25 % et doit être en mesure d'en produire les justifications.

Autres critères

Il peut exister d'autres critères évolutifs retenus par des organismes particuliers et bien définis :

- Le Pouvoir Propagateur du Feu(arrêté du 28 août 1959 de la Marine Marchande) a pour but de tester la vitesse de propagation superficielle pour un matériau en l'état ou revêtu (peinture, revêtement de mur, de sol) soumis à l'action d'un rayonnement calorifique.

  Les classes retenues dans cet essai sont :

  - la classe 0, lorsqu'il n'y a ni inflammation, ni ignition
  - les classes 1, 2, 3 et 4, dans lesquelles les vitesses moyennes de propagation, les limites de propagation et éventuellement les temps d'inflammation sont progressivement croissants.

- Opacité des fumées : des essais ont été effectués récemment en France par le Syndicat Général des Fabricants de Panneaux dérivés du bois auprès du LNE (Laboratoire National d'essais) à la chambre N.B.S.

Ces essais ont permis de vérifier un excellent comportement en la matière, puisque les fumées dégagées sont des fumées claires moins opaques que celles dégagées par le pin naturel, et beaucoup moins opaques que celles dégagées par le chêne naturel.

Ces études se poursuivent en étroite collaboration avec les Instituts et Organismes scientifiques français et étrangers.

1.2 Réglementation française en matière d'incendie

Cette réglementation très détaillée, est actuellement établie pour :

- les Etablissements recevant du public
- les Immeubles de Grande Hauteur (I.G.H.)
- les Bâtiments d'habitation
- les Etablissements industriels et commerciaux

La plus grande partie de cette réglementation émane du Ministère de l'Intérieur bien que d'autres instances officielles aient édicté des règles particulières (Ministère du Travail). Les Commissions consultatives de sécurité sont chargées de veiller au respect de cette réglementation. Elle comporte dans chaque domaine :

- des règles générales
- des règles particulières propres à certains type de bâtiments

Nous ne retiendrons ici que les règles générales concernant les parois verticales :

Etablissements recevant du public

Eléments de construction	Résistance au feu	Réaction au feu	Textes Réf.	Observations
Eléments porteurs ou autoporteurs constituant le gros oeuvre du bâtiment	stable au feu 1/2 h à 1 h 1/2		CO 14	
Façades : - en rez-de-chaussée - en étage P/H inférieur à 0,8 P/H supérieur ou égal à 0,8 - Façades vitrées		 M2 M2 M3	CO 19	P = distance minimale des constructions en vis-à-vis occupés par des tiers H - hauteur de la construction la plus élevée régles spéciales en fonction de la masse combustibles au m2
Encloisonnement des escaliers et ascenceurs	coupe-feu 1 h 1/2 ou pare-flammes 2 h	M3 ou M2	CO 22 CO 25 et Education Nationale	
Cloisons - intérieures de distribution - limitant les couloirs de circulation ou les locaux destinés au sommeil	 pare-flammes 1/2 h coupe-feu 1 h	M3 ou M2	CO 29 et Education Nationale *	

Eléments de construction	Résistance au feu	Réaction au feu	Textes Réf.	Observations
Eléments de décoration en relief Revêtements (décoratifs, insonores aux autres) en panneaux de particules		M1	CO 30 et Education Nationale	
- sur paroi incombustible ou si vide, entre panneau et paroi, bourré par un matériau incombustible		M3 ou M2		
- éloignés d'au maximum 5 cm des parois mais avec traverses formant cellules closes dont la plus grande dimension est inférieure à 3 m.		M2		
- éloignés des parois sans recoupement		M1		
Complétés par les régles * particulières d'utilisation dans les locaux scolaires apportées par le Ministère de l'Education Nationale en 1974.				Voir page 13. Ch. 2-34

Immeubles de Grande Hauteur (I.G.H.)

Eléments de construction	Résistance au feu	Réaction au feu	Potentiel calorifique (kg de bois par m2 de surface en oeuvre	Textes Réf.	Observations
- matériaux de construction en contact avec l'air (ou rapidement en contact en cas de feu)		M2		GH 13	M3 possible pour portes, parquets collés en bois et revêtements de sols collés
			15*	GH 14	
- revêtements muraux dans halls d'entrée				GH 13	collés sur supports coupe-feu 2 h.
- éléments mobiliers			25*	GH 49	

* Les 15 kg comprennent en moyenne par compartiment, l'ensemble des matériaux combustibles, les revêtements de sols collés étant déduits.

Bâtiments d'Habitation

Eléments de construction	Résistance au feu	Réaction au feu	Potentiel calorifique (kg de bois par m2)	Textes Réf.	Observations
Eléments porteurs verticaux - 1re famille & 2e famille à rez-de-chaussée en bande - 2e famille (autres bâtiments)	 stable au feu 1/4 h. stable au feu 1/2 h.			art. 6	
Face externe des parois extérieurs verticales - 1re famille - 1re famille (si à plus de 4m. de la limite de parcelle) - 2e famille - 3e et 4e familles a) rez-de-chaussée b) étages D/H inférieur à 0,8 D/H supérieur ou égale à 0,8		 M3 M4 M3 M2 M2 M3		 art. 7 art. 7 art. 7 art. 8	 en 3e & 4e familles le panneau de particules peut être utilisé en habillage D: distance minimale de l'immeuble en vis-à-vis H: hauteur la plus élevée de ces deux immeubles
Panneaux de façades vitrées 1) - 3e famille C + D > 0,60 m. - 4e famille C + D > 0,80 m. 2) - 3e famille C + D > 0,80 m. 4e famille C + D > 1,00 m. 3) - 3e famille C + D > 1,30 m.			 1,5 1,5 à 5 sup.	art. 9	en 3e famille seulement sous certaines conditions C & D voir croquis
Parois des cages d'escaliers 2e famille 3e famille - paliers & volée d'escaliers	 pare-flammes 1/2 h. pare-flammes 1 h. stable au feu 1 h.			art. 12	non exigibles dans les maisons individuelles en bandes à deux niveaux
Parois séparant les celliers - groupés en étage & ouvrant sur des dégagements communs des autres parties de l'immeuble	coupe-feu 1 h.			art. 13	

Sécurité contre les risques d'incendie et de panique dans les bâtiments relevant du Ministère de l'Education Nationale

Compte tenu des conditions particulières d'utilisation des locaux scolaires, le Ministère de l'Education Nationale en France a apporté en 1974 des exigences complémentaires aux règles générales de sécurité concernant les locaux recevant du public.

A titre d'exemple, pour les cloisons industrialisées combustibles, et en complément des prescriptions de l'article CO 29, les règles énoncées ci-dessous sont appliquées :

a) Pour les cloisons simples réalisées par un seul panneau :

- avoir un revêtement difficilement inflammable (peinture, tissus)
- être au moins difficilement inflammable, M2
- si des vides existent, ils doivent être comblés avec des matériaux incombustibles, MO

b) Pour les cloisons avec vide d'air réalisées par assemblage de deux panneaux non inflammables ou difficilement inflammables sur bâti moyennement inflammable au moins

- avoir un revêtement difficilement inflammable (peinture, tissus)
- si les panneaux sont non inflammables M1, aucun recoupement n'est nécessaire
- si les panneaux sont difficilement inflammables M2, le vide doit être recoupé dans les conditions définies au paragraphe 2 de l'article CO.31, principalement en parties haute et basse, et il est procédé à un bourrage tel que prévu au paragraphe 4 de l'article CO.31

c) Pour la qualité des joints

- vérifier que les joints verticaux et horizontaux entre panneaux et entre panneaux et planchers sont prévus en conformité avec ceux décrits au procès-verbal de classement de résistance au feu des panneaux de cloisons utilisés.

2. LES MATERIAUX DERIVES DU BOIS

Parmi les nombreux matériaux issus de la matière première bois formant les parois verticales de bâtiments propres à l'activité humaine (habitat, commerce, industrie, bureaux, éducation, loisirs) on distingue cinq grandes familles :

- les panneaux contreplaqués et lattés
- les panneaux de particules
- les panneaux de fibre
- les panneaux décoratifs issus des trois précédents
- les panneaux composites associant l'un de ces types à d'autres matériaux, tels que mousses, fibres minérales, plastiques, etc.

Les panneaux contreplaqués et lattés, forment le secteur industriel le plus ancien parmi les dérivés du bois, et celui qui a été le plus étroitement associé au développement du béton armé, en dalles, banche ou précontraint.

Les contreplaqués dits "de coffrage" touchent directement le gros oeuvre de bâtiment ou de génie civil, au niveau du chantier, mais ne font pas partie intégrante de l'ouvrage. A ce titre, ils ne sont pas concernés par les problèmes de feu.

Par contre, les contreplaqués dit "extérieurs", fabriqués et vendus en France sous la marque de qualité "extérieur CTB-X" d'épaisseur 10 à 22 mm, jouent un rôle important dans la conception même des façades, en tant qu'éléments de structure (parois porteuses) ou éléments de remplissage (panneaux d'allèges). A ce titre, ils sont soumis aux règles de sécurité portant sur les façades, et la limitation des effets de propagation du feu. Aussi, les fabricants proposent-ils des contreplaqués ignifugés à coeur ou par traitements appropriés de surface, classements en réaction au feu :

- M3 - Moyennement inflammable pour épaisseurs 14 mm
- M2 - Difficilement inflammable

Les panneaux contreplaqués courants, d'usages intérieurs sont en général de faible épaisseur (4 à 10 mm); ils sont également proposés dans ces différents classements pour répondre aux exigences requises en agencement ou décoration intérieure des bâtiments d'habitation ou des locaux publics. Celles-ci peuvent aller jusqu'au classement M1 non inflammable.

Les lattés par leur composition même sont fabriqués en général dans les épaisseurs supérieures à 15 mm et de ce fait classés M3 par nature.

Les panneaux de particules représentent le secteur à progression la plus rapide et plus spécialement dans le domaine de l'agencement. Son prix, ses formats, ses cadences industrielles, en font un matériau privilégié d'aménagement intérieur, de ce fait, il participe de plus en plus à la structure même du bâtiment.

A ce titre, les panneaux de particules sont directement visés par les règles de sécurité incendie sous les divers aspects :

- réaction au feu
- résistance au feu
- pouvoir calorifique
- toxicité et opacité des fumées

Les efforts des industriels se sont portés sur tous ces points et ont été d'autant plus fructueux que leur technologie de fabrication se prête aux multiples possibilités de combinaisons performantes en association avec les produits chimiques, minéraux ou autres.

Ainsi sont nés, en réponse aux exigences des pouvoirs publics :

- les panneaux ignifugés à coeur M2 ou M1 d'usage courant ou CTB-P à replaquer pour milieux secs

- les panneaux hydrofuges CTB-H et ignifugés à coeur M2 ou M1 pour équipement sanitaire ou doublage de murs
- les cloisons coupe-feu-pare-flamme 1/2 heure, 1 heure, 1 heure 30, selon les épaisseurs et en leur état
- les mêmes cloisons mais également ignifugées à coeur et classées M2
- les portes coupe-feu-pare-flamme 1/4 d'heure ou 1/2 heure à âme en panneaux de particules
- les mêmes matériaux enduits ou prépeints sans affectation du classement en réaction au feu d'origine

Les panneaux de fibre dont la progression est moins spectaculaire, restent largement utilisés en parois verticales soit en revêtement pour les fibres dures, soit en matériau de remplissage isolant, pour les panneaux isolants imprégnés ou non de produits anti-fongiques.

Ces derniers, pour leurs caractéristiques d'absorption phonique, doivent rester apparents et à ce titre, présenter des classements en réaction au feu M1 ou M2, selon leurs domaines d'emplois. Des traitements de surface appropriés leur permettent cette conformité.

Leur emploi comme matériau de remplissage a été limité en France, par l'obligation de n'utiliser que des matériaux incombustibles (M0), comme nous le verrons ci-après pour les panneaux composites.

Pour les panneaux de fibre durs traditionnels, fabriqués par voie humide, le classement M2 peut-être obtenu par procédé d'imprégnation.

Pour les panneaux de fibre durs, et grâce au procédé de fabrication "à sec", le traitement dans la masse, permet d'obtenir des panneaux ignifugés M2 en épaisseurs de 3 à 6 mm.

Comme pour les panneaux contreplaqués, et les panneaux de particules ignifugés à coeur, grâce à la durabilité dans le temps du traitement dans la masse en cours de fabrication, la validité du classement obtenu en réaction au feu est égale à 10 ans.

Les panneaux décoratifs : ce secteur plus récent touche aux aspects les plus nobles des matériaux de base précédents.

Ceux-ci servant de supports, sont ennoblis :

- soit par une simple finition d'apprêt telle que vernis, cire, produit d'imprégnation, lorsque le parement de bois naturel reste apparent et notamment dans le cas des contreplaqués essences fines ou des panneaux de particules replaqués bois
- soit par une stratification directe par films imprégnés de résines synthétiques sur supports panneaux de particules ou panneaux de fibre durs
- soit par contrecollage de revêtements divers souples ou rigides tels que stratifiés, lamifiés, feuilles de métal, films PVC, toiles de textiles, etc.

- soit par enduction apprêt peintures ou laques sous cylindres, pistoletteuses ou machines rideaux.

Toutes ces opérations comportent trois critères essentiels :

- compatibilité avec les supports concernés
- durabilité (comportement dans le temps)
- comportement au feu

Les deux premiers critères physico-chimiques étant supposés réglés, le critère de comportement au feu reste une obligation majeure pour cette famille de matériaux.

Le revêtement superficiel, généralement le premier en contact avec la flamme ou soumis à élévation de température, peut présenter un risque important de propagation, ou d'intoxication.

Le report de ces opérations en usine, permet de mieux maîtriser la qualité d'exécution et de garantir le classement final en réaction au feu par le contrôle du produit fini.

Le choix des produits de finition, ou de leur procédé de mise en oeuvre permet désormais à la fabrication de panneaux décoratifs bois naturel, ou de finitions diverses sur supports dérivés du bois dans les classements M3, M2 et M1.

Les <u>panneaux composites</u> : De techniques plus récentes, les panneaux composites se développent à la mesure du manque de main-d'oeuvre spécialisée sur les chantiers. Ils permettent de construire "à sec" sans apport d'humidité, et en toutes saisons des parois séparatives intérieures ou des parois de façades, en éléments complets. L'enchaînement des corps de métier se fait suivant un planning déterminé et sans aléas.

Les panneaux dérivés du bois, par leurs possibilités d'adaptation aux différentes fonctions, leurs facilités d'usinage (sciage, clouage, vissage, défonçage) et de collage, leur facilité de transport, sont associés de façon privilégiée aux matériaux d'isolation thermique ou phonique.

Leur conception doit répondre à des règles précises, complexes et conjointes, exemple :

- résistances mécaniques
- amovibilité
- réaction au feu
- résistance au feu
- isolation phonique

3. <u>LES DOMAINES D'APPLICATION ET LEURS EXIGENCES</u>

Tout en restant dans le domaine des parois verticales, et en nous limitant aux emplois les plus généraux, nous pouvons distinguer :

- les parois de distributions intérieures :
  - menuisées, sur ossature bois
  - simples
  - composites
- les parois extérieures
- les parois de doublage de mur (ou habitallage) :
  - doublage de murs extérieurs
  - doublage de murs intérieurs

Problèmes propres à chaque paroi sous l'aspect feu : les exigences propres aux parois verticales sont directement liées à la fonction remplie par l'élément concerné en cas d'incendie, vis-à-vis :

- de l'ouvrage en général (stabilité des structures)
- des éléments mitoyens (stabilité de la paroi et liaisons avec les murs de façades, de refend, etc.)
- des ouvrants (portes, fenêtres)
- des occupants ou des sauveteurs (sécurité civile)
- des biens (assureurs)

A titre d'exemple : Pour les parois menuisées sur ossature bois, les panneaux sont fixés de façon solidaire ou amovible sur le site. L'ossature doit remplir sa fonction propre de portance et de stabilité au feu. Les panneaux jouent un rôle d'écran dont la durée coupe-feu-pare-flamme peut être testée en laboratoire officiel indépendamment du rôle porteur de l'ossature, et associée ou non à des matériaux de remplissage. De telles parois sont généralement classées coupe-feu-pare-flamme 1/2 heure avec deux parements en panneaux de particules de 12.

coupe-feu-pare-flamme 1 heure avec deux parements en panneaux de particules de 22 mm.

Ceci, sous réserve d'un soin particulier au droit des joints verticaux et horizontaux.

Pour les parois simples, généralement d'épaisseurs supérieures à 22 mm, et très couramment de 50 mm, les panneaux, d'ordinaire en particules de bois, jouent à eux seuls le rôle de cloison.

Convenablement assemblés entre eux, ces éléments obtiennent des classements coupe-feu, pare-flamme : 1/2 heure en 22 ou 25 mm
1 heure en 50 mm

Leur rigidité permet l'amovibilité par systèmes de clips, vérins, mais suppose alors des précautions particulières de montage pour maintenance de ces classements.

Pour certains ouvrages et à partir de 50 mm un rôle porteur peut-être requis en complément.

Dans les parois composites, les deux fonctions précédentes se combinent soit par une ossature plus légère, soit par contrecollage, ou mise en place de matériaux complémentaires pour des raisons thermiques, phoniques, etc.

Ces cloisons gardent le bénéfice de l'amovibilité, et peuvent être élaborées en usine, permettant ainsi les combinaisons les plus diverses, sous réserve que l'élément de remplissage, s'il n'existe pas de barres transversales, soit incombustible (M0) lorsqu'il s'agit de bâtiments scolaires.

Dans les trois cas, et selon les exigences ci-après les produits dérivés du bois seront choisis en fonction de leur classement en réaction au feu : M3, M2 ou M1.

Pour les parois de doublage, sur murs de maçonnerie, les exigences de sécurité incendie portent essentiellement sur les systèmes de fixation et tout en tenant compte du phénomène de parois froides (risques de condensation), attirent l'attention sur le vide et son contenu entre panneau et mur.

4. CONCLUSION

Par la qualité et la diversité de ses productions, l'industrie française des panneaux dérivés du bois s'adapte aux exigences de la réglementation en matière de sécurité contre l'incendie.

Quels qu'en soient les niveaux, à l'exception des I.G.H. (limitation de pouvoir calorifique), le bois, sous ses formes les plus diverses :

- contreplaqués
- panneaux de particules
- panneaux de fibres
- panneaux décoratifs
- éléments composites

reste l'un des matériaux les plus sécurisants de notre cadre de vie. Il faut cependant souhaiter qu'une harmonisation des méthodes d'essais et critères de classement, permette aux industriels et utilisateurs européens de parler le même langage, tout en donnant aux législateurs les moyens de préparer l'harmonisation des règlementations communautaires.

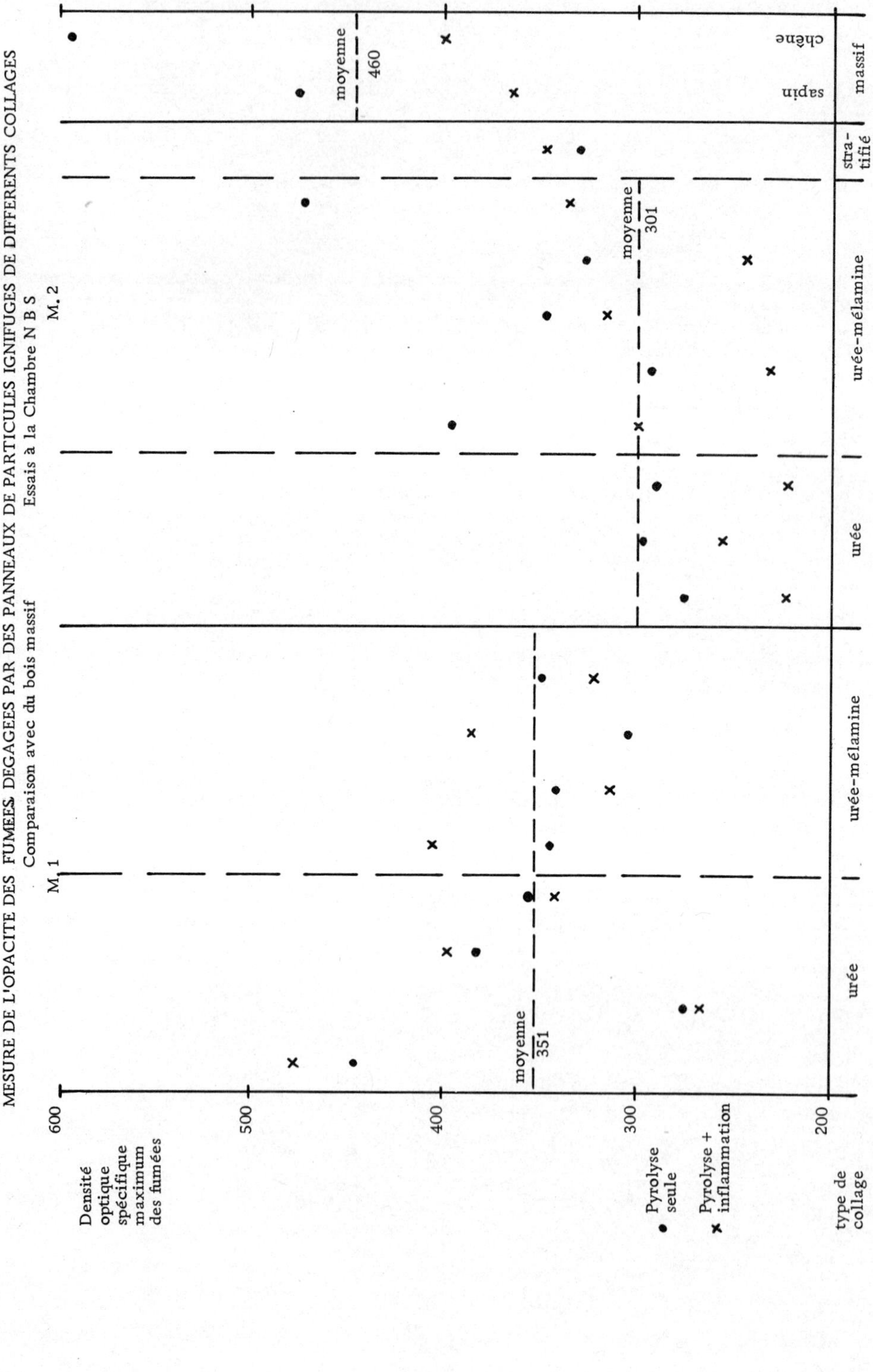
MESURE DE L'OPACITE DES FUMEES DEGAGEES PAR DES PANNEAUX DE PARTICULES IGNIFUGES DE DIFFERENTS COLLAGES
Comparaison avec du bois massif
Essais à la Chambre N B S
M. 1
M. 2
Densité optique spécifique maximum des fumées
600
500
400
300
200
moyenne 351
moyenne 301
moyenne 460
Pyrolyse seule
Pyrolyse + inflammation
type de collage
urée
urée-mélamine
urée
urée-mélamine
stra-tifié
sapin
chêne
massif

Behaviour of Wood Structures and Components (Beams and Columns) in Fire

B. Compin
Institut de Recherches Appliquées au Bois (IRABOIS)

SUMMARY

Because wood in inflammable, there have long been doubts, so far as fire safety is concerned, regarding its use as a building material.

It is only in recent years that it has been recognized that, in certain conditions, the behaviour of wood in fire is not more dangerous than that of other traditional building materials.

FRENCH FIRE SAFETY REGULATIONS

France has quite comprehensive fire safety regulations. The set of regulations applying to buildings used by the public is the one which is most relevant to the subject under consideration, because it prescribes performance rather than the materials which may be used.

Thus, the regulations in question do not specify the type of building materials to be used and therefore allow the use of wood in certain conditions, which depend on its behaviour in fire.

Roughly speaking, the framework of such buildings must have a one-half hours fire resistance rating. Provided that the building components are "moderately inflammable" (category M3), however, such resistance is not required in single-storey buildings in which the parts of the frame are visible from the ground or when such parts are separated from areas of the building used by the public by a fire-proof floor or counter floor which can resist fire for one-half hour.

Thus, this regulation allows the use of wood as a structural component.

EXPERIMENTS IN FRANCE

Few experiments have been made, although IRABOIS has carried out tests on solid wood and glued laminated wood columns and beams.

These tests not only confirmed the results obtained in other countries, but also enabled IRABOIS to prove that the behaviour of wood in fire is very satisfactory and even quite spectacular.

IRABOIS invited representatives of trade journals, newspapers and television, designers, insurers and fire safety departments to observe these tests.

EXPERIENCE GAINED FROM FIRES IN WOOD STRUCTURES

A few years ago, several fires occurred in buildings made of glued laminated wood.

Most of these fires broke out during construction.

In each case, the fires lasted at least about one-half hour.

The fact that there have been no cases in which structures collapsed shows that wood structures perform well in cases of fire.

THE POSITION ADOPTED BY INSURANCE COMPANIES

In France, the General Assembly of Fire Insurance Companies (Assemblée Plénière des Sociétés d'Assurances contre l'Incendie" (APSAI) is composed of insurance companies specializing in the coverage of industrial fire risks.

This Assembly bases its premiums on a schedule in which building materials are divided into three categories, namely, hard, medium-light and light materials.

Wood used in building structures is classified as a "hard material" by insurance companies when it meets certain requirements of the C.S.1 schedule of specifications.

According to these specifications, arch and cross-beam frame components made of glued laminated wood must have cross-sections of at least 15 x 15 cm and 11 x 30 cm.

Wood structures built according to these specifications are not subject to higher fire insurance premiums than similar masonry structures.

Thus, fire safety regulations in France allow the use of wood in construction. It should also be noted that insurers do not object to the use of this building material, subject to certain requirements, which are not difficult to satisfy.

The possibility of using wood in construction has been allowed because it has been possible to prove that wood performs well in case of fire and its performance is recognized both by insurers and by safety commissions.

Comportement au Feu des Constructions et Eléments en Bois

B. Compin
Institut de Recherches Appliquées au Bois (IRABOIS) (France)

POUTRES ET COLONNES

Le bois, en tant que matériau de construction, a longtemps fait l'objet de réserve en raison de sa combustibilité.

Déjà au Moyen Age, il était interdit de laisser apparentes les ossatures des maisons en pan de bois. C'est ainsi par exemple que dans le centre de la ville de Rouen, dont les maisons étaient pour la plupart construites selon cette technique, les façades avaient été plâtrées pour répondre aux prescriptions d'un édit. Ce n'est que récemment que Rouen a procédé au déplâtrage de ses façades.

Il en est de même à Paris où un grand nombre de maisons avaient une ossature de bois ainsi que dans bon nombre de villes de province.

Il est vrai qu'à cette époque, les moyens de prévention, et surtout de sauvetage, étaient quasiment inexistants. Un incendie pouvait prendre une ampleur catastrophique en détruisant des quartiers entiers.

En outre, les connaissances concernant l'évolution d'un incendie étaient à peu près nulles.

L'utilisation du bois en charpente, depuis le début du siècle, s'était amoindrie en France. Cette réduction résultait de l'apparition de nouveaux matériaux de construction : fonte et acier et surtout béton armé. Et puis, au cours des deux dernières décennies, l'emploi du bois pour la réalisation des structures s'est à nouveau considérablement développé pour permettre à la charpente bois de retrouver la place qu'elle n'aurait jamais dû céder. Ce renouveau de la construction bois est dû à l'apparition de nouvelles techniques et tout particulièrement celle du bois lamellé collé qui a permis la réalisation d'ouvrages spectaculaires de très grande portée tel le hall d'expositions des foires de Tours qui couvre 100 mètres par 100 mètres, sans point d'appui intermédiaire.

Ces réalisations ont été possibles parce que les préjugés à l'égard du comportement du bois sous l'action de l'incendie sont tombés avec le temps et avec l'acquisition de meilleures connaissances.

LA REGLEMENTATION FRANCAISE EN MATIERE DE SECURITE INCENDIE

La nécessité d'une réglementation en matière d'incendie a été ressentie à la suite de catastrophes telle celle du Bazar de la Charité en 1897 qui fit 130 morts. C'est le Préfet Lépine qui créa la Commission Supérieure des Théâtres, chargée d'établir un règlement pour les établissements de divers ordres, dans le département de la Seine.

En 1927, une ordonnance fut prise concernant les précautions à observer dans les établissements de spectacles, à Paris. C'est dans cette dernière réglementation que l'on trouve le fondement du règlement de sécurité des établissements recevant du public.

Cette réglementation est intéressante en ce sens qu'elle n'est pas descriptive, mais exigentielle et qu'elle ne fait pas de distinction selon la nature des matériaux.

Elle permet donc, dans certaines limites, fonction de son comportement au feu, l'emploi du bois dans la construction.

En gros et très schématiquement, les pièces de charpente doivent être "stables au feu $\frac{1}{2}$ heure". Toutefois, sous réserve que les éléments constructifs soient au moins "moyennement inflammables" (classés M3). Ce comportement n'est pas exigé dans les bâtiments à simple rez-de-chaussée dans lesquels les pièces de charpente sont visibles du sol ou lorsque ces pièces sont séparées des locaux accessibles au public par un plancher ou faux plancher coupe feu $\frac{1}{2}$ heure.

Ainsi, cette réglementation permet l'emploi du bois dans les structures.

Une réglementation existe également pour les bâtiments autres que ceux recevant du public.

C'est en particulier le cas des bâtiments d'habitation. Ceux-ci sont classés en quatre familles.

La première famille concerne les habitations individuelles isolées ou jumelées ou de deux niveaux au plus.

La deuxième famille comprend les habitations individuelles isolées ou jumelées à plus de deux niveaux et les habitations collectives dont le dernier niveau habitable est situé à moins de 8 mètres au-dessus du sol.

La troisième famille comprend les habitations dont le plancher bas du logement le plus haut est situé à moins de 28 mètres du sol utilement accessibles aux engins de secours.

Enfin, la quatrième famille englobe les habitations dont le même plancher est situé entre 28 et 50 mètres au-dessus du sol.

Pour ces constructions, il est exigé que les éléments porteurs verticaux soient stables au feu :

- 1/4 d'heure pour la première famille
- 1/2 heure pour la deuxième famille
- 1 heure pour la troisième famille
- 1 heure $\frac{1}{2}$ pour la quatrième famille.

Quant aux planchers, ils doivent respectivement être coupe-feu 1/4 d'heure (1ère famille), 1/2 heure (2ème famille), 1 heure (3ème famille), 1 heure $\frac{1}{2}$ (4ème famille).

En ce qui concerne les ossatures bois, ce sont essentiellement les exigences des première et deuxième familles qui retiennent notre attention et l'on peut facilement se rendre compte que celles-ci peuvent y répondre.

Evidemment, cette réglementation impose d'autres prescriptions mais il n'est pas dans notre propos de l'analyser dans son détail.

La réglementation française vise également les immeubles de grande hauteur. Le principe qui a guidé le législateur est de subdiviser l'ouvrage en compartiments. Si un sinistre survient dans un tel compartiment, il ne doit pas se propager aux autres et les structures doivent être en mesure de conserver une stabilité suffisante.

Dans cette réglementation, les potentiels calorifiques sont limités. Le bois est visé au premier chef puisque l'on indique la correspondance en kg de bois du potentiel calorifique limite autorisé (2,25MJ/m2) soit environ 15 kg de bois. En outre, l'emploi de matériaux moyennement (M3) ou facilement (M4) inflammables est interdit sauf en ce qui concerne les portes et les parquets collés en bois. Il est vrai qu'il est assez peu vraisemblable que soient édifiés des immeubles de plus de 28 mètres de hauteur en ossature bois.

Ceci constitue un tour d'horizon extrêmement succinct de la réglementation incendie française, car il ne peut être question, dans cette étude, de l'étudier dans son détail.

Nous n'avons mentionné, ici, que les textes essentiels. D'autres prescriptions existent, dispersées dans de nombreux textes, comme par exemple le règlement sanitaire départemental type ou encore, les magasins généraux, voire même le code du travail.

Nous mentionnerons toutefois la législation des Etablissements Classés, c'est-à-dire des établissements industriels. Celle-ci est ancienne. Les premiers arrêtés types datent du début du siècle et jusqu'au début de l'année s'appuyaient sur une loi de 1914.

Cette législation est descriptive mais elle est en cours de modification pour devenir, en ce qui concerne le feu, exigentielle comme les autres textes législatifs que nous avons mentionnés.

Ce tour d'horizon, très rapide, de la réglementation française en matière de sécurité incendie montre qu'elle n'exclut pas l'emploi de structures en bois moyennant le respect de certaines spécifications qui ne se montrent pas abusives.

EXPERIMENTATIONS FAITES EN FRANCE

Des essais de résistance au feu d'éléments de structures en bois ont été réalisés un peu partout dans le monde. Ils tendent tous vers des résultats assez convergents.

Pour notre part, nous avons fait exécuter des essais sur des poteaux et des poutres en bois assemblés, en bois massif et en bois lamellé collé.

Ces essais ont confirmé les résultats obtenus à l'étranger mais surtout, ils nous ont permis de prouver que le comportement du bois sous l'action du feu était très satisfaisant et même spectaculaire.

Nous avons d'ailleurs donné à certains de ces essais, une publicité en conviant à y assister des représentants de la presse professionnelle, de la grande presse écrite et télévisée, des concepteurs, des services de sécurité et surtout des assureurs.

Les essais ont été confiés au laboratoire du feu du Centre Scientifique et Technique du Bâtiment.

Ils ont donc confirmé ce à quoi l'on pouvait s'attendre, à savoir que les pièces composées sont plus sensibles à l'action du feu que les pièces massives, et que les pièces en bois lamellé collé avaient un comportement tout à fait favorable. La forme des sections qui peut être appréciée par le rapport surface exposée/volume présente une très grande importance sur les durées de résistance au feu.

Ainsi, deux poteaux présentaient des rapports surface/volume respectivement égaux à 0,29 dm^{-1} et 0,37 dm^{-1}. Le second poteau, bien que pesant 37% de plus que le premier a résisté moitié moins de temps que le premier.

Il convient donc essentiellement, du point de vue résistance au feu, de rechercher des sections, en évitant celles qui paraîtraient intéressantes du point de vue résistance des matériaux et économie de matière. Il convient d'éviter de faire des profils.

Il résulte de l'ensemble des essais qui ont été effectués que, immédiatement en-dessous de la couche de bois détruite par le feu, le matériau demeure intact et conserve intégralement l'ensemble de ses qualités.

L'humidité que contient le bois joue aussi un rôle primordial.

On sait que sous l'action de la chaleur, il se produit à l'intérieur du bois une migration de l'eau vers les parties les moins chaudes. On s'en aperçoit nettement lorsqu'on effectue des essais de cloisons ou de portes. De larges taches d'humidité apparaissent dès le début de l'essai sur les faces non exposées.

Dans le cas de poutres ou de poteaux, les éléments sont attaqués sur les quatre faces. L'eau se trouve alors rejettée dans les parties centrales des pièces.

D'autre part, la chaleur de vaporisation de l'eau est très élevée. Des thermocouples introduits à l'intérieur des échantillons testés à différentes profondeurs confirment ce fait. Les températures marquent un palier très net à 100°C alors que la température ambiante atteint 900°C.

Ce phénomène explique le bon comportement des bois lamellés collés et surtout le fait qu'il ne se produit pas de décollement.

Les expériences montrent que la vitesse de destruction en profondeur du bois est de l'ordre de 6 à 7/10 mm par minute pour un bois résineux courant (sapin, épicéa) et qu'elle est pratiquement indépendante de la température du four.

Cette progression de l'attaque serait un peu plus élevée dans le cas de portes ou cloisons où une seule face est exposée au feu.

Il est probable que l'une des raisons en est que, dans le cas d'éléments exposés au feu sur les quatre faces, l'eau, ne pouvant pas s'éliminer par la face non exposée, migre beaucoup moins facilement et absorbe par conséquent un nombre sensiblement plus élevé de calories pour son évaporation par la face exposée.

En définitive, tout le monde s'accorde sur ce chiffre de 6 à 7/10 de mm/mn, et nous avons des raisons de penser qu'il est exact.

Nous avons, en effet, fait réaliser un essai de résistance au feu d'une poutre caisson en contreplaqué.

Nous recherchions un degré de stabilité au feu d'une demi heure et nous avons dimensionné les poutres en nous appuyant sur les chiffres cités.

Ces poutres avaient les caractéristiques suivantes :

- longueur : 5,00 m
- hauteur : 0,70 m
- épaisseur totale : 179 mm
- membrure haute et basse : 150 x 75 mm
- treillis : 2 f : 37 x 100 mm.

L'ensemble habillé sur chaque face d'un contreplaqué de 15 mm d'épaisseur, agrafé.

Ce contreplaqué ainsi que la totalité des bois mis en oeuvre n'avaient subi aucun traitement.

Les deux poutres ont été essayées ensemble, dans un même four, par la station d'essais au feu du Centre Technique Industriel de la Construction Métallique de Maizières-les-Metz, selon le programme normalisé.

Elles ont résisté 33 minutes, obtenant ainsi le classement recherché "stable au feu $\frac{1}{2}$ heure".

Des études ont également été menées sur des bois protégés.

La protection des bois par imprégnation à coeur, intéressante du point de vue de l'amélioration de la réaction au feu, présente par contre certains inconvénients dont les principaux sont l'augmentation sensible du prix et la réduction de certains avantages inhérents au bois, notamment en ce qui concerne l'aspect et les facilités de travail, par suite des teneurs élevées des produits minéraux incorporés. De plus, il n'est pas prouvé que la résistance au feu soit beaucoup modifiée par de tels traitements.

En ce qui concerne les enductions, les avantages très réels dans le cas de produits de bonne qualité, du point de vue de la réaction au feu (retard par exemple de l'inflammation de l'ordre de 8 à 10 minutes) sont beaucoup plus faibles, quant à la résistance au feu, pour des bois de fortes sections. Le retard à la pénétration est souvent de l'ordre de 3 à 5 minutes. Dans le cas d'une vitesse de destruction de 0,6 mm à la minute, le gain correspondant serait ainsi de 5 à 8 minutes. En tablant sur un retard de cinq minutes, on obtiendrait un résultat équivalent en majorant le côté d'un poteau de section carrée de 1 cm environ. Le prix de revient serait sans doute moins élevé, compte tenu du prix du produit et de l'application et l'on éviterait les sujétions relatives à la fragilité de la couche enductrice.

Si l'on veut obtenir des résultats intéressants, il faut faire appel à des enduits épais dont l'un des plus traditionnels et des plus efficaces est le plâtre; mais l'on masque alors l'aspect du bois.

ENSEIGNEMENTS TIRES DE QUELQUES INCENDIES DE BATIMENTS A OSSATURE EN BOIS

Nous avons connu il y a quelques années plusieurs incendies qui se sont déclarés dans des bâtiments à structure en bois.

La majorité de ceux-ci ont pris naissance alors que la construction n'était pas terminée.

Ce sont souvent les étancheurs qui ont été à l'origine de ces sinistres. Une gamelle de bitume chaud se renverse sur la toiture, le bitume s'enflamme, coule dans les gouttières et descentes d'eau pluviales et propage le feu dans tout le bâtiment.

Ce sont aussi les utilisateurs de chalumeau qui mettent le feu à un bardage en bois fraîchement verni.

Dans chaque cas, le feu a sévi pendant près ou plus d'une $\frac{1}{2}$ heure et les structures ne se sont jamais écroulées.

Evidemment, il a été nécessaire de les remplacer car les bois étaient carbonisés et inacceptables esthétiquement.

Nous citerons le cas d'une chaufferie provisoire d'un groupe d'immeubles.

Cette chaufferie était implantée sous un dôme circulaire en charpente bois lamellé collé. La couverture était constituée de bardeaux d'asphalte posés sur des panneaux de particules de 22 mm d'épaisseur.

La chaufferie comportait trois ouvertures : deux portes diamétralement opposées et un lanterneau circulaire de deux mètres de diamètre, au faîte de la coupole.

Dans la nuit, du 3 au 4 janvier 1973, à la suite d'une fuite de fuel, un incendie se déclara à l'intérieur de la chaufferie. Les pompiers interviennent environ 25 minutes après le début du sinistre et se rendent maîtres du feu après plus d'une heure.

L'ouvrage ne s'est pas affaissé et est resté étanche à l'eau.

La chaufferie constituant un bâtiment provisoire, il a été jugé inutile de la reconstruire après ce sinistre. Elle est toujours en place.

Nous citerons un dernier exemple : l'incendie des Hydrocarbures de Saint Denis, le 29 janvier 1968, d'origine malveillante.

Cet incendie est le plus violent que les pompiers de Paris aient eu à combattre durant les 50 dernières années.

Les circonstances ont voulu qu'une charpente en bois, moderne, triangulée, se soit trouvée en bordure de ce gigantesque brasier.

Les trois halls principaux qui constituaient cet ensemble industriel étaient distants d'environ 5 mètres des premiers réservoirs de stockage de carburant. La conception des bâtiments était assez classique : poteaux en béton porteurs, remplissage en parpaings, combles en charpente bois couverts en plaques d'amiante ciment ondulé, avec interposition de plaques en verre ondulé armé, partie supérieure des pignons en plaques d'amiante ciment sur ossature bois, la partie basse étant constituée du mur limitatif en maçonnerie sur une hauteur de 4 mètres.

Cette charpente fut donc soumise à une épreuve qu'il serait absolument impossible de réaliser en laboratoire. Il est très difficile de se faire une idée sur les températures atteintes. Dès le début du sinistre, lorsque les pompiers arrivèrent, ils ne pouvaient s'approcher à moins de 200 mètres du brasier, tant la chaleur rayonnée était intense.

Lorsque le Général des pompiers sut que la charpente de cette usine était en bois, il fit donner l'ordre à ses hommes de pénétrer dans ces locaux et à certains de prendre poste sur le toit.

LA POSITION DES ASSUREURS

Les Sociétés d'Assurances qui, en France, prennent en charge les risques découlant d'incendie dans l'industrie, sont regroupées au sein de l' "Assemblée Plénière des Sociétés d'Assurances contre l'Incendie" A.P.S.A.I.

Cette Assemblée a établi un tarif dans lequel les matériaux constitutifs des constructions industrielles sont classés en trois catégories : les matériaux durs, semi-légers et légers.

Les matériaux durs et semi-légers sont nommément désignés dans des listes limitatives. Tous les autres matériaux sont considérés comme légers.

Les taux de prime sont établis, entre autre, en fonction des proportions de matériaux légers, semi-légers et durs qui entrent dans les constructions.

Il convient de préciser que le tarif établi par l'A.P.S.A.I. ne s' applique en théorie qu'aux bâtiments industriels. Mais en fait, il sert bien souvent aux assureurs pour établir les primes afférant à des bâtiments d' autres destinations.

Les assureurs classent "matériau dur" le bois entrant dans la composition des structures lorsqu'il répond à un certain nombre de spécifications contenues dans un Cahier de spécifications techniques dit "Cahier C.S.1.".

Ces spécifications fixent, entre autres, les sections minimales suivantes :

- Pour les constructions en bois massif :
  - Eléments verticaux comprimés (poteaux) 15 x 15 cm

- Eléments complémentaires (poteaux de bardage, lisses et liens de contreventement) 8 x 18 cm
- Eléments d'entretoisement : 5 x 15 cm

- Pour les constructions à ossature en bois lamellé collé :

- Elément d'ossature principal 11 x 30 cm.

Les sections minimales des éléments complémentaires et d'entretoisement, qu'ils soient en bois massif ou en lamellé collé, étant les mêmes (8 x 18 et 5 x 15 cm) que pour les ossatures en bois massif.

Les primes d'assurances incendie des constructions industrielles en bois répondant à ces spécifications, assez peu contraignantes d'ailleurs, ne subissent pas de majoration.

CONCLUSION

Les règlements de sécurité français autorisent l'utilisation du bois pour la réalisation moyennant certaines prescriptions.

Les assureurs de leur côté, ne sont pas opposés à l'emploi de ce matériau moyennant le respect de certaines conditions.

Ceci reflète le bon comportement du bois sous l'action du feu à condition cependant que celui-ci soit mis en oeuvre en section massive.

Design Concepts for Fire-resisting Constructions

J.G. Sunley, MSc, FIStructE, FIWSc.
Director, Timber Research and Development Association,
United Kingdom

INTRODUCTION

In this paper, I wish to make two main points. If we are to make progress and obtain as quickly as possible the best balance between safety and economy in our structures designed against fire we must not shackle ourselves to a regulatory system which has not got a rational base. In this context the two points I wish to make are firstly a sensible combination of theoretical design and properly planned testing will permit more logical progress than either separately. Secondly, informed people using judgement based on adequate research etc. will ensure more rapid progress than a rigid regulatory system.

If we compare structural engineering design for structural safety and design against fire we can detect many differences. The progress that has been made in structural design could be a guide on how to deal with fire design.

Structural engineering design emerged primarily as a theoretical excercise with little or no recourse to test in the early days. Structural members and components have been theoretically designed for a hundred years or more and considerable experience exists on the performance of these designs. In more recent years more effort has been directed towards prototype testing to provide confirmation of previous designs. Also the fact that designs have existed and failures have taken place enable calibration to be carried out between methods of design and real life performance.

The design of fire resisting structures has been based on different principles. There has been very little theoretical design with nearly complete reliance on testing. It has to be appreciated that errors due to difficulties in simulation of real-life loading and other effects make the results less dependable than is usually appreciated. Aircraft design which arrived later than structural design has probably been based on a better combination of theoretical design and test than either structural members in buildings or design of fire resisting structures. This occurred since it is essential to optimise weight of the structure and the penalty of too conservative design is readily apparent.

AIMS OF DESIGN

Generally safety in design is concerned with two main functions. One is the avoidance of loss of life and the other is the economic loss of property.

Sometimes these two effects may be considered together, at other times separately. For example, a structure which either collapses or contains a major fire will be both a danger to life and suffer economic loss with regard to the property itself. However, with regard to fire design it could be considered desirable to spend more of the cost of the building on means of escape rather than on protecting the building. In that context more attention would have been paid to loss of life than loss of property.

Unfortunately, regulations have tended to confuse the difference between loss of property and loss of life by being in theory concerned with loss of life and in practice probably just as concerned with loss of property. The aim of design is therefore the achievement of acceptable probabilities that the structure being designed will not become unfit for the use for which it was required during its life and that it will offer reasonable safety to the occupants etc.

This means that to achieve these aims, the method should be based on scientific theories, experimental data and experience of past design practice interpreted as logically as possible by informed people. It must also be emphasised that safety is not simply a function of design and test calculations, it also depends on the quality of control exercised in fabrication and the supervision on site.

The maintenance of the condition of use of the structure and full maintenance of the structure itself during its intended life are also implicitly assumed.

The failure of a structure, either structurally or by fire may cause injuries and deaths to people, disrupt social life and bring personal suffering. These consequences cannot be quantified in pure economic terms and consideration has to be given to what society is willing to tolerate.

It has to be accepted that most natural activities concerned with people and buildings involve risk of unintended and unwanted occurrences and fire design is no exception. The levels of risk which society will tolerate are influenced by many factors, some real, some emotional, some imaginary. Some of the factors involved are:

i) The historic background of the risk. This indicates why Mr. Malhotra in his paper drew attention to the fact that many major changes in regulation follow immediately on some major disaster. The emotional content of such decisions is apparent.

ii) The difficulties of controlling the risk. Various human activities have to go on or we wish them to go on and even if the difficulties of control are immense and the risks great society will accept the risk.

iii) The degree to which the risk is imposed or is voluntary. Where children congregate in large numbers in schools etc. a higher level of safety is demanded than for example if they are being looked after personally in their own home or even in a motor car.

iv) The nature of the consequences. Here again the emotion of loss of life and injury by fire is probably greater than any other type of accident. Therefore a lower risk of this happening is demanded than virtually any other.

v) The personal or social benefits for which the risk is suffered. People will obviously take a greater risk for their own personal gain than if occupied in gain for a third party. This means that people in a factory or a hospital or school where people congregate demand lower risk than a one-man self-employed business.

There is less tolerance of risk in any event which will have severe consequences for a large number of people rather than to a whole series of small events. Thus one single incident involving say a hundred fatalities will be considered much more serious than a hundred occurrences, each with one fatality. This factor is directly related to the sensational value of such events to the news media and in the political sphere.

ECONOMIC CRITERIA

The experience of loss of life due to fire is obviously minute in proportion to other everyday hazards (e.g. road accidents). However, the aversion to such loss through fire is largely emotive and does not pay due regard to the economic consequences of maintaining a disproportionate amount of safety in a fire risk compared to any other risks. Theoretically, any selection of risk associated with fire should be made rationally with due consideration of the financial cost to society since in the absence of unlimited resources any increase in expenditure on construction to reduce risk of fire must imply a reduction in expenditure on other needs.

A wrong balance in degree of safety and economic consequence can sometimes be counterproductive. This was evidenced in the aftermath of the Fairfield Home enquiry in which many authorities in the United Kingdom redirected funds from providing additional homes to the provision of elaborate alarm equipment in existing homes. Therefore a large number of elderly people remained at high risk in unsuitable accommodation to ensure the provision made for others was above criticism. There is thus a case for examining the likely economic consequences of human requirements even though these are very emotive.

In many instances the direct economic consequence of failure by fire are of paramount importance and should govern the design targets. Fortunately because of the "insurance situation" in fire where human life is not involved this is relatively easily achievable. This arises because fairly good statistics are available for loss by fire and insurers in a competitive situation are obliged to provide sensible cover at reasonable cost. The treatment of this aspect of fire design is more easily quantified than the emotional one of death or injury. Although it is common to divorce the social and economic factors they are interlinked and should wherever possible be considered together. Although greater material well-being has brought with it a higher valuation of life and a greater demand for security it still has to be paid for and regardless of the effort and money expended there can never be absolute safety. It is a fact that increased expenditure on safety is subject to diminishing returns. It has to be faced that the safer any structure is made inevitably the more expensive it will become, and with limited economic resources they will then be able to afford fewer structures.

We cannot, therefore, avoid the consideration of probability of risk and economic consequence. Also at the same time continuous education should encourage people to aim for similar risks rather than be too emotional about particular ones.

FIRE RESISTANCE

Fire resistance implies that the efforts to restrict development of the fire (flame spread, rate of heat release etc.) have failed and the situation within an enclosure is out of control. This means that it is highly desirable that the fire should be contained while the rest of the building continues to stand and carry out its normal functions. At the same time, this will, of course, limit the risk to human life. This means that within the context of design for fire resistance we are not dealing with surface characteristics of wood based materials. As explained earlier, in structural design for normal structural safety, buildings are usually designed as a series of elements which are then interconnected so that the whole building acts together and benefits from the various interactions.

Originally, in structural design, no advantage was claimed from the interactions between components which in practice add considerably to the strength and stiffness of the whole. However, more recently this question has been looked into, and advantages identified so that more economic structures can result from better knowledge of the saftey of the whole.

LESSONS TO BE LEARNED FROM STRUCTURAL ENGINEERING

Although structural engineering is now fully codified and regulated in Building Regulations etc. it had a long period of development in which the only control was through Codes of Practice which were purely advisory documents. This permitted a freer development of an art which has consequently reached a higher level of development than design for fire resistance. It could be unfortunate that the regulatory systems have taken over fire control at a much earlier stage in its development. We could therefore well suffer from a lack of freedom in design by informed people with guidance from Codes of Practice and continue to be fettered by a strict regulatory system.

I would contend that design against fire is little different from that against "normal" structural collapse except that the state of the knowledge and testing technology is less advanced. For example, the functioning of the whole in fire resistance is assumed on the basis of tests on the parts. It is only recently that even simple joints between elements have been evaluated and very few tests on this situation have been conducted so far.

COMPARTMENTATION

The main design concept in fire resistance is compartmentation. In semi-detached houses the party wall makes each house a compartment. Generally, compartments are recognised by the Building Regulations, are large, and great care is exercised to ensure that they remain firetight. There is, however, considerable merit from a property damage point of view in scaling down compartmentation so that individual rooms are compartments in a loose sense. This looseness can be tolerated since most fires are effectively fought by the fire brigade who can cope with minor leaks. Thus even an ordinary hollow core interior door will hold back a fully developed fire for a useful period, particularly if water is sprayed on the back, as will tend to happen in normal fire fighting operations. The point is that in every element with a containment function there is a useful degree of fire resistance even if not required by regulations. Thus although there may be no regulations requirement, bedroom ceilings are frequently plasterboard and should be so rather than a non-fire-resisting foam.

Although fire engineering design is in its infancy compared with other aspects of structural design there is information available on material properties. For example, the predictability of timber in fire is better than in normal structural strength properties. There also appears to be less variability between individual components tested against fire than with similar components tested structurally. This means that the same principles as have evolved in structural design can be applied to fire resistance design.

CODE FOR FIRE RESISTANCE DESIGN

A design code could therefore be written for fire resistance which:

i) provides a method of calculation for such design

ii) provides for validation by prototype testing

iii) deals with widely used traditional solutions by means of "deemed to satisfy" designs.

Until recently within many countries only ii) and iii) have been acceptable solutions.

A method of design calculation requires:

a) Definition of a fire to be resisted. This is done in ISO DIS 834, an international Standard which has been widely adopted as national Standards.

b) The known reaction of timber products and ancillary ones to the above type of exposure.

The problem of fire resistance design can be looked at in two situations:-

i) where there is enough sacrificial timber to allow charring for the period of fire resistance required by the particular building component and still retain sufficient residual strength, stiffness and insulation to perform its original function in loadbearing use, e.g. beams, columns etc.

ii) where protection is required but all or some of the timber is not exposed for the full fire resistance period required. This would apply particularly to walls where for example the initial exposure is to the lining material.

For beams, columns, it is frequently more economic to increase their dimensions to provide sacrificial timber rather than use protection of other materials. In fact, timber can be used to protect steel beams and columns in an analogous sacrificial way. However, for stud walls, linings usually form the first line of defence. The same might apply to certain joisted constructions.

Although there is primary protection provided by the normal linings, secondary protection can be given in both cases from the insulating materials such as rock- and slag-wool, although they are not necessarily put there primarily for that purpose. Even after destruction of the first and secondary defences the loadbearing timber itself still has some fire resistance, although this is, of course, similarly finite. Using the three components, elements can be designed with adequate fire resistance in addition to other properties such as economy, sound resistance, appearance, durability, etc.

LOADING

The question of the loads to be considered in a fire situation has been discussed under limit state design considerations. In the UK each year, approximately one building in twelve hundred is affected by a fire which damages the structure (as opposed to the contents). This is approximately one building in 500 000 per day. The probability that this particular building on that particular day will be subject to maximum snow and wind loads is extremely remote indeed. Therefore, one may be justified in designing the actual maximum design dead applied load and ignore the specific effects of wind, snow etc. on the fire affected structure. The stability during the rebuilding of the damaged building could be the critical requirement.

WHOLE BUILDING DESIGN

A Code could give advice from which a competent designer or consultant could design a whole building by putting the elements together. Thus a plasterboard lining could be assumed to have a certain fire life proportional to thickness. Rock wool insulation could also have a value. The fire resistance of studs could be ascertained from charring rates. A summation of the above could give a predicted fire resistance for the combined structure. The Table in the attached Appendix gives proposed values in a draft British Code for predictive design.

In addition to involving load interactions, other points to be considered are the behaviour of junctions and physical/mechanical interactions. There is already some test evidence that commonly used and well-designed junction details between floors and walls do not constitute weaknesses with regard to containment of fire. Junction design of this nature usually involve practical forms of construction which avoid through gaps or exploitable gaps even with the settlement of components and deflection under loading etc. It can be desirable to restrict the size of cavities especially if ventilated by blocking or mineral wool gaskets etc. The main point is that it is possible to design logically in this way rather than depend on testing.

APPENDIX

An Illustrative table of Extracts from the Draft of a Method of Predictive Design for Fire Resistance

INTERNAL WALLS REQUIRED TO RESIST FIRE FROM ONE SIDE AT A TIME - SINGLE LEAF WALLS - JOIST, STUDS AT 600mm CENTRES UNLESS OTHERWISE STATED

Material	Min. thickness mm	Min. density kg/m³	Contribution to fire resistance: stablity mins.	Integrity mins.	Insulation mins.	Restrictions
Linings, exposed Plasterboard to BS1230	9.5 12.5 19.0 etc.		15 20 30	12 17 25	12 17 25	Nailing at 150mm centres with 30mm galvanised nails 40mm " " 60mm " " — Tapered edge boards with exposed joints reinforced with paper tape and filled with Gypsum based filler or 5mm Gypsum plaster on suitable boards (joints scrimmed) — Max. stud centre 450mm for 9.5mm
Wood chipboard	12.0 etc.	600	14	12	12	Nailing at 300mm centres at perimeter and 400mm at intermediate supports 30mm nails
Med. density hard-board to BS1142	9.0 12.0	600 600	10 17	10 15	10 15	Max. nailing centres 250mm for 9mm. Nailing at 100mm centres at board perim. and 200mm centres at intermed. supp. with 40mm nails
Timber Studs Softwood (except Western red cedar)	38.0 44.0		5 10	- -	- -	Carrying up to maximum design load according to CP112
Softwood (except Western red cedar)	38.0 etc.		10	-	-	Carrying up to 50% design load according toCP112
Cavity Insulation None	-	-	0	0	0	Fixed to protect sides of studs: blankets or slabs
Mineral wool to BS3958 (excluding those manuf'd from glass)	25.0 50.0 75.0	48 48 48	10 12 15	10 12 15	10 12 15	See Fig. 5a, b
Linings, unexposed None	-	-	-10	-10	-10	
As exposed linings with symmetrical constructions						
Plasterboard to BS1230	9.5 12.5 etc.		0 0	8 13	8 13	Nailing at 150mm centres with 30mm galvanised nails 40mm " " — Tapered edge boards with exposed joints reinforced with paper tape and filled with Gypsum based filler or 5mm Gypsum plaster on suitable boards (joints scrimmed)
Wood chipboard	12.0	600	0	8	8	Nailing at 300mm centres at perimeter and 400mm at intermediate supports 30mm nails
Med. density hard-board to BS1142	12.0	600	0	10	10	Nailing at 100mm centres at board perim. and 200mm centres at intermed. supp. with 40mm nails

Design of Timber Structures for Fire Resistance in Finland

Esko Mononen
Chief Engineer, Ministry of Interior, Finland

INTRODUCTION

Design rules for fire resistance of glued laminated beams and columns were specified in Finnish fire regulations in January 1969. These rules were based on rather extensive fire tests and the mean charring rate in standard fires for spruce and pine was given as

charring (mm) = 34 t, where t is the time of fire in hours.

Rounding rules for corners were also given. It was assumed that the strength in bending and compression will drop 25% and the modulus of elasticity 15% during fire. The loads used were usual design loads for permissible stress design except snow load, which was assumed to diminish during 2 1/4 hours. Later, in 1972, the concept of using reduced imposed loads in fire was introduced and was also used in tests.

In recent years, the "limit states design" approach has been in preparation and is already in use for concrete structures. A draft for fire resistance of load bearing structures and another for limit states design of timber structures has also been sent for public comment. Joint Nordic recommendations for structural safety are still under preparation and will also affect the design for fire resistance.

LOADS AND SAFETY FACTORS

In the latest Finnish draft for fire resistance of load bearing structures, the standard fire is considered as an accidental load. In combination with fire the imposed load for floors in dwellings, hotel rooms, offices, schools and hospital wards is given as 0.75 kN/m^2 and in restaurants, assembly halls, shops, dancehalls and theatres as 2.00 kN/m^2. Snow and wind loads are reduced to 30% of design load. This means that snow load is reduced to 30% of its maximum 30-year value and wind load to 30% of its maximum 50-year value (velocity pressure of 3 sec. gust). Snow and wind loads are not supposed to act simultaneously in combination with fire. It could be said approximately that wind and snow loads are lower than these reduced loads about 80% of time. When a fire is not taken into account the load combinations for roofs are snow load + 50% of wind load and wind load combined with a 50% snow load. A bigger reduction during fire has been considered justified because a fire is a more rare phenomenon than e.g. ultimate snow load (characteristic load multiplied by a partial load factor) and the fire load is supposed to last usually only 0.5-2 hours. In limit states design all loads during fire are multiplied by a load factor 1.0 as for serviceability limit states.

DESIGN VALUES FOR TIMBER

The following values are based on the latest draft for fire resistance of load bearing structures.

For glued laminated structures of pine or spruce glued with phenol-formaldehyde or resorcinol-formaldehyde resin adhesives the charring rate (in mm) is calculated as 0.6 x t, where t is the time of fire (in minutes). The radius of rounding in corners is calculated as 10 + 0.3 t (millimeters). According to tests these values are valid at least for a standard fire of 120 minutes. Longer tests are not yet performed in Finland for wood beams. Minimum breadths of cross sections are 70 mm for 30 minutes fire resistance, 140 mm for 60 minutes and 280 mm for 120 minutes.

For solid members the charring rate is assumed as 0.8 t because of splitting.

The strength values of the residual section are taken without any reduction as limit states values (5% fractile) for short term loadings. The load factor for material values (dry values of the residual section) is taken as 1.0.

Unless a beam does not collapse earlier, an ultimate limit state for safety is reached when the deflection is one thirtieth of the clear span. The used value of the modulus of elasticity is the mean value as for serviceability limit states.

RECENT PRACTICE

Load bearing glued laminated timber structures can be used in one or two-storeyed buildings. The exposed sides of structures are untreated but between the beams or columns the interior surfaces usually have to fulfil requirements for ignitability and flame-spread. In tests the charring rate for fire-retardant treated wood has been greater than for untreated wood, and therefore only the latter is in use. Metallic parts in joints and hangers are protected with wood, mineral wool or gypsum slabs if exposed parts can affect the fire resistance.

In some cases a design concept is used where a beam is acting with an unprotected tension rod for normal snow load and without the tension rod in fire with reduced snow load.

Glued laminated beams and columns are used to a large extent for fire-resistive structures. The calculation of fire resistance is considered by engineers as reliable as for protected steel structures and even easier.

Harmonization of Building Regulations: Other Fire-linked Factors

The Effect of United States Building Code Fire Regulations on Wood Construction

Roderick B. Buchan
Assistant Vice President, National Forest Products Association, USA

There is no such thing as a fireproof building. The contents of most buildings will burn and it is those contents, not the building's construction, which pose the gravest threat to life and property. Yet building codes in the United States continue to over-emphasize the significance of non-combustible structural elements and tend to prohibit or seriously restrict use of wood in specific cases. On the other hand, steel and concrete are accepted by the codes as conforming to the "non-combustible" definition with little regard for the other adverse effects of fire on these materials.

The important consideration in judging fire safety of a structure is how well the building performs in protecting life and property in a fire rather than the composition of the structural material used. As that view becomes more widely acknowledged by the various regulatory agencies, the opportunity for wood use in the United States will increase. This paper will examine the US Building Codes, how they operate, how they inhibit wood use, and what the American wood industry is doing about it.

THE BUILDING CODES

Early Fire Regulations

Before the current century, wood construction in buildings was practically unrestricted in the United States. But as cities grew in size and number, public concern over fire and structural hazards also grew.

In the early 1900s, local authorities began to restrict use of wood construction. Soon competing materials interests recognized the market value of codes and standards which prohibited wood use. It became clear to lumbermen of that period that without a programme to counter these piecemeal challenges, many wood uses would be regulated out of existence or rendered uncompetitive. Since 1918, protecting wood industry interests in building codes and ordinances has been a major responsibility of the National Forest Products Association, the members of which include other wood associations as well as forest product companies.

Model Codes

There are over 8300 municipalities and 1500 counties in the United States that have a law regulating building construction for the purpose of assuring the safety, health and general welfare of the people in the community.

Thirty states have regulations on manufactured housing. Effectively monitoring and influencing these local building codes would be economically impractical if it were not for the existence of the national model building codes, which are developed by private agencies and adopted by individual local and state jurisdictions.

The five principal model building codes in the United States are: the Standard Building Code, published by the Southern Building Code Congress, Int., widely used in the South; the Uniform Building Code, published by the International Conference of Building Officials, primarily used in the West; the Basic Building Code, published by Building Officials and Code Administrators, Inc., widely used in the Midwest and Northeast; the National Building Code, published by the American Insurance Association and used in many small jurisdictions throughout the US; and the One and Two Family Dwelling Code, published by all the model code groups and used throughout the US. Most of the local and state codes in the country are similar to or adaptations of these, with the exception of large cities such as New York, Los Angeles, Philadelphia, and Chicago.

Fire Related Provisions

Unlike insurance requirements, these codes when adopted by the state or municipality, have the authority of law. All building designs must meet code specifications for maximum permissible heights and floor areas. These specifications are based on certain characteristics of the building, including fire zone location; type of occupancy (hospital, office, residence, etc.); construction materials and systems; setbacks from property lines; exits; and automatic fire suppression systems.

Federal Regulations

A large number of federal agencies also develop and enforce regulations which must be complied with in addition to local and state codes. Notable among these federal specifications are the Department of Housing and Urban Development standards (HUD) which regulate one and two family dwellings and multi-family dwellings where federal loans are involved, and all mobile homes, and the Department of Health, Education, and Welfare standards which regulate most health care facilities and institutional occupancies. Federal regulations are having an increasingly significant effect on all building construction, particularly as tightened requirements relating to life and fire safety are promulgated.

Fire Marshal Rules

Further, local and state Fire Marshal Rules and Regulations affect building construction. These requirements can take precedence over local building codes in public occupancies such as auditoriums, schools, and hospitals. The typical fire marshal tends to be overly protective in matters of fire safety. Fire marshal rules in a number of states have more restrictive controls on construction than the applicable building codes.

RESTRICTIONS ON WOOD

Naturally, all these regulations have a direct bearing on how wood can be used and whether it can be used economically. Wood construction is prohibited for many uses, such as large assembly buildings, mental hospitals, and some types of schools.

Interior Finish

Since wood is a basic finishing material, code requirements for interior finish greatly affect wood use. The surface flammability of finish materials is regulated by flame spread ratings which usually are divided into three classes. Untreated wood falls into the more restricted class and is severely limited under most codes for use as finish in certain rooms and stairway shafts, corridors and exitways in all buildings except single family dwellings and accessory buildings.

Wood paneling, doors and trim can be specifically prohibited in certain occupancies such as large assembly areas and hospitals. This prohibition is rationalized on the basis that the life hazard is high and therefore combustible materials should not be allowed. Some jurisdictions place such severe limitations on flame spread rates on interior finish that wood is prohibited for all practical purposes. In some jurisdictions interior finish requirements have been applied to built-in furniture installations in an effort to prevent use of wood. Even wood cabinets in high-rise apartment buildings have been prohibited on the basis of interior finish requirements.

Roof Coverings

Another area where the codes in many instances are overly limiting is the regulation of roof coverings. Wood shingles and shakes are in strong demand for use on residences, restaurants, churches, and offices, not only for appearance and durability, but, increasingly, for their insulation value. However, many jurisdictions continue to prohibit use of wood roofs for any purpose.

Windows

High-rise buildings have proven a very favorable market for wood windows; and, again, with energy conservation of prime concern, the superior insulation characteristics of wood windows encourage their use. With appropriate design procedure and proper setback from property lines, wood windows are not a fire hazard. Nevertheless, many codes have a blanket provision prohibiting such uses.

Fire Resistance

Historically, most one and two family residences with appropriate clearances from property lines have had no fire restrictive requirements. In recent years, however, federal agencies have initiated regulations which require fractional fire ratings (i.e. less than one hour fire resistance) for structural components of one and two family residences. Appropriate fire restrictive requirements based on performance are desirable. However, some of the proposed limits place wood framing in an unfavorable competitive position.

Noncombustible Materials

In some instances "noncombustible" materials have structural advantages. Generally, however, there is no demonstrated reason to require "noncombustible" materials to achieve fire and life safety. Unfortunately, code writing authorities continue to specify use of such materials. To enable wood to be used where "noncombustible" construction is required, fire retardant treated wood, pressure treated with salts to inhibit ignition and spread of flame, was developed. Fire tests conducted on fire retardant treated wood (FRTW) such as the tunnel test, corner test, heat release tests and full scale fire endurance tests show that FRTW can perform as well or better than noncombustible material, particularly light gauge metal products. Although a substantial market for FRTW has been developed in the US, codes still require noncombustible materials in many instances where treated wood can do the job.

Exiting

Many codes permit "noncombustible" buildings to have greater distances to exits than buildings which use wood in their construction, although required exitways are the same for all types of construction. This design feature can eliminate wood construction in some instances.

Fire Zones

The number and size of fire zones in a city can also unnecessarily restrict timber construction. The American Insurance Association, which underwrites major insurance companies in the US, recognizes the economic danger of a too-restrictive system of fire zones and generally advocates a single fire district to cover high-value congested commercial areas. Fire zones that are too numerous or too large often curtail new construction and restrict remodeling. This leads to economic deterioration of these areas. Although the practice of assigning <u>all</u> commercial and industrial property to a single zone is not considered good nor effective fire zoning, many jurisdictions still do it.

Requirements within fire zones have a considerable effect on wood construction. The most restrictive fire zones prohibit use of all wood construction and reduce allowable floor areas for wood and masonry buildings. Other fire zones, and some jurisdictions have as many as four, restrict use of wood to some degree. For example, they may prohibit use of wood roofs, wood siding, wood eaves and overhangs, and wood canopies.

Fire Suppression Systems

The enviable performance record of automatic fire suppression systems has not thus far been fully recognized in the US, particularly when used with wood frame construction. Most codes do not permit adequate "trade-offs" from fire resistive requirements to make use of fire suppression systems economically feasible.

Rate of Heat Release Theory

One of the most significant current issues in the building and fire code area involves the concept of the rate of heat release of a material as opposed to its potential heat in relationship to the fire resistance of construction materials and assemblies. There is a trend in the United States toward limiting allowable potential heat of construction materials or assemblies, particularly in large buildings which offer opportunity for use of heavy timber construction and fire retardant treated wood. The American wood products industry takes the view that potential heat that is not released or which is released at a slow rate is of little consequence in a fire. The superior fire performance of large timber members compared with the rapid failure of unprotected steel members under similar fire exposure points up the fallacy of the potential heat approach.

Comparative fire tests on wood and steel beams have provided ample evidence to refute the popular belief that any material which will not burn automatically will endure exposure to fire.

In addition, a history of more than 150 years of excellent performance under severe fire exposure has established the superior fire endurance of heavy timber framing. For many years, heavy timber framing has been considered a preferred risk by fire insurance rating bureaux and a superior type by fire protection authorities.

NEW PROBLEM AREAS RESTRICTING WOOD

A number of ominous building regulations have appeared on the scene which can have a severe impact on current markets for wood products. These are concerned with smoke development, toxicity of products of combustion, flame spread of floor coverings and restrictions on use of furniture, shelves, cabinets and other furnishings.

Smoke Development

Even though most authorities in the fire field agree that the tunnel test, developed primarily to determine flame spread of wood materials, is not an appropriate test method for measurement of smoke, it has been widely referenced in codes throughout the US for that purpose. Some smoke development limits in these codes prohibit use of wood construction. The National Bureau of Standards recently developed a smoke measurement test method which, while recognized only for research and development purposes, is being referenced by some enforcing authorities, particularly federal agencies, in their construction standards. The limits set in many instances eliminate wood products. These prohibitions are being enforced despite the fact there has been no documented information to show a correlation between test results and smoke development under actual fire conditions.

Smoke Toxicity

Historically in the US, codes that regulated toxicity of combustion products have done so by stating that products of combustion shall be no more toxic than that developed by the burning of untreated wood, thus offering no problem to wood products. However, the increasing use of new plastic materials in buildings made it necessary to develop standards for the toxicity of products of combustion. Initial development work has shown that current regulations may not adequately control the smoke and toxic gases emanating from these new materials. Thus, the present trend is to remove from the code all reference to toxicity until an appropriate test method can be developed. This removes from the code the favorable position that wood once enjoyed with regard to toxic combustion products. A new test standard is likely in the near future which appears certain to contain limits that will eliminate some uses of wood.

Floor Coverings

With the wide use of carpeting for floor coverings and some unfortunate fire experiences attributed to spread of fire due to carpeting, the development of floor covering tests was inevitable. The widely known "pill test" is required of most carpeting used in the US and appears to have eliminated carpets that were "bad actors". However, new floor covering tests have been developed. Foremost in this area is the "Radiant Panel Test" developed by the National Bureau of Standards. This test method has been referenced in federal specifications and its application limits will preclude use of wood flooring in some instances even though there is no documented evidence to indicate that wood floor surfaces are a factor in spread of fire.

Furnishings

An area of great concern is the potential effect of test programs presently underway for development of standards to regulate furniture, cabinets, shelves, and other furnishings. While the thrust of these tests will be toward regulation of plastic materials, the "limits" set by regulatory authorities will likely restrict, if not eliminate, many traditional wood products uses.

Fire Flow Concept

Another new concept in building regulation which has surfaced in recent years is the "fire flow" concept. This concept regulates building size based on the amount of water supply available to suppress the fire. Despite knowledge that most fires originate in contents and that severity is directly related to contents, the "fire flow" concept places little emphasis on these facts. It allows significantly larger areas for unprotected metal buildings than for wood frame buildings, all on the basis of less fuel contribution. These restrictions are imposed despite the fact that under the same fire conditions and with the same contents the structural integrity of a metal building would not be as great as a wood frame building. Some jurisdictions in the US already have incorporated this concept into their building codes.

Field Services

Still another development affecting the wood products industry is use of survey data to support proposals for greater restrictions on use of wood products. Recently, a survey of fires in residential structures in a very limited geographic area was funded by the federal government. Results of the survey were interpreted as suggesting new code requirements were necessary on the use of wood siding, wood balconies, and wood roof overhangs. Unless offsetting statistical data can be developed which involves heavy cost or other solutions can be found, a significant market for wood products will be lost.

WOOD INDUSTRY EDUCATIONAL AND RESEARCH PROGRAMS

The need for developing highly credible technical performance data for materials and methods of construction is becoming increasingly important as the government regulatory process becomes more complex and more sophisticated. Governmental initiatives in fire protection and in related wood product areas, such as energy conservation and product liability, have placed major new responsibilities on those in the building fields. New federal agencies such as the Consumer Product Safety Commission and the Energy Research and Development Administration have been established to oversee development of standards which cover these emerging areas of public protection. Education of the building official in these areas has become a high priority activity of parent code-writing bodies. The wood products industry is intimately involved in this educational process.

Since fire performance traditionally has been the most severe area of code restriction on use of wood, a major portion of the National Forest Products Association's research and data development effort is in this field. The wood industry is involved in a basic fire research program to overcome the most restrictive of the fire performance limitations on wood use - that of combustibility.

In a special project which is now in its fourth year, the industry is supporting research being conducted by a fire research specialist at the US Department of Commerce National Bureau of Standards. The purpose of this effort is twofold: first, to demonstrate that it is the rate at which a material burns and gives off heat and toxic gases that determines its contribution to fire severity; and second, to develop fire experience data which demonstrate that the structure of the building is not significant in determining the extent of fire hazard when reasonable and appropriate fire and life safety features are observed.

This is a long-range program and progress has required painstaking development of equipment and repetitive testing of many materials. Ultimately, standards for testing and test limits will have to be established before the benefits of this important concept can be proposed for inclusion in the building codes. The acceptance of code regulations based on the rate-of-heat release concept, however, should overcome a longstanding impediment to wood use - that of its combustibility.

The wood industry also is sponsoring a Graduate Research Fellow at the University of Utah to investigate the area of smoke and toxic gases in a fire. The Flammability Research Center at the University of Utah is one of the country's foremost institutions for such work. This research is highly significant because little is known about the products of combustion of wood even though smoke and toxic gases are recognized as major contributors to loss of life in fires.

The forest products industry also participates in the American Society for Testing and Materials' and the National Fire Protection Association's fire research activities. These groups develop the consensus standards by which many wood products are tested and rated for fire safety. The wood industry's involvement with them is essential to insure a balanced view of the problems of fire testing.

It is not the intent of this paper to suggest that wood materials should be permitted without exception. Appropriate regulations are necessary to life and fire safety and the orderly growth of a community. The goal of the industry is to recommend fire safe design of buildings, regardless of the material used. This can best be achieved through establishment of realistic fire hazard requirements and by application of meaningful fire ratings to structural elements.

This paper has attempted to establish the fact that all wood products used in building construction are regulated to a substantial degree in the United States and that the restrictions are being intensified.

Canadian regulations governing use of wood products are quite similar to those in the US. In recent years there has been significant growth in exchange of information among the regulatory authorities of the two nations. It can be safely stated that regulatory problems affecting use of wood products in the US will "rub-off" on Canada and vice versa.

If wood construction is to fulfil its expanding role in providing safe housing as well as efficient commercial and industrial structures, all segments of the wood products industry throughout the world should particpate in development of the test methods and standards used to regulate such construction, and seek appropriate and equitable regulations that maintain and assure the markets for this renewable natural resource.

Building Regulations Concerned with Fire Safety and their Impact on the Use of Wood Products in the Nordic Countries

A. Mårtenson
Chairman of the Sub-Committee for fire protection of NKB

Building regulations, as concerns fire protection, aim at limiting damage by fire to life, health, and property. To this effect regulations, more or less completely, deal with measures to provide

a) prevention of fire
b) escape routes (number, siting, design etc.)
c) adequate fire stability of building frames
d) limitation of the spread of fire and smoke
e) installations for fire-fighting and rescue.

The impact of the regulations on the use of wood products is due to the fact that such products are combustible and some of them vigorously so. This calls for restrictions to keep the risk of fire damage to a reasonable level. At least in Sweden and other Nordic countries, however, wood products are in no way discriminated against, compared to other products of a combustible nature. This can be illustrated as follows.

To prevent the outbreak of fire (a) above), for instance, it is prescribed that building components of combustible materials such as wood, either be kept beyond certain safety distances from fire-places, chimneys and the like, or be protected by a suitable lining. To this kind of regulations there are virtually no divergences that cannot be solved by proper testing.

In escape routes (b) above) the use of combustible material is still more restricted and care is taken not only to combustibility as such but also to smoke development properties. Hence wood products are not allowed as ceiling or wall lining in escape routes, unless they are so treated as to comply with relevant test demands. According to the difficulties encountered in drawing adequate conclusions from small-scale laboratory tests, especially for thermoplastic products, the authorities in many cases have found it necessary to base their decisions on ad hoc room-scale tests for which there is unfortunately no proper test method available.

Securing a minimum fire stability (c) above) for load-carrying structures traditionally calls for specified walls, columns, slabs etc., to be made of non-combustible material - confining the use of wood products to a rather limited amount. This is based on the experience that - although grading results from standard fire tests may be equally good - structures of conventional stone materials such as concrete, brickwork etc. have a superior residual integrity after the test, compared to combustible structures. Such an integrity is apparently of great value in many cases where fire-fighting may be delayed, e.g. because rescue operations are given priority, or out of necessity limited, in order to keep the fire within the compartment of origin. For the present, regulations of this kind are only partially based on systematic analysis, which should comprise both the need and the value of said residual integrity. Further information on this field might well be of interest for wood products.

To prevent a fire from spreading (d) above) the regulations operate with compartmentation but also with restrictions on ceiling and wall linings although these are less rigorous than for escape routes. As far as compartmentation is concerned there are wide possibilities to use wood products, provided that the test requirements are met. As for linings, the possibilities are somewhat more limited but still real enough at least on walls and especially in single-family houses (where wood panelling is accepted also in ceilings) and other relatively low buildings.

These outlines can only give a broad view of the conditions within the Nordic countries and naturally less for other countries. However, it is safe to state that there is room and need for co-ordination in building regulations in this and many other fields. Yet much and increasing work is being carried out to improve this situation. One sub-regional organization dealing with such matters is the Nordic Committee on Building Regulations (NKB), established by the national building authorities in Denmark, Finland, Iceland, Norway, and Sweden. The purpose of its activities is the co-ordination of the technical building regulations in the countries involved.

The object of these efforts is to harmonize requirements and methods of verification into a system that is neutral in relation to the national administrative systems in the various countries. Although concerned with such regulations of all kinds, the Committee has lately given high priority to a few fields, among them fire protection. To begin with its aim is, as soon as possible, to achieve co-ordination on the basis of current regulations, without waiting for inaccessible results from present or future research.

As of now (January 1977) the following has been achieved:

1) a set of guidelines for structural fire protection regulations for single-family houses (established 1975); and

2) a draft system of inter-Nordic fire technical classes (under final revision)

With regard to the use of wood products these guidelines (1) above) - as reported to the ECE seminar on industrial production and use of wood-based products in the building industry at Sandefjord in June 1975 - imply that:

a) as concerns acceptable linings, reference is made to the properties of solid wood, due to known deficiencies in available test methods;

b) the minimum fire stability or resistance, where required, is generally 30 minutes, in accordance with a standard fire test, which is easily achieved by the use of wood products; and

c) a joint wall separating houses built together is to have 60 minutes fire resistance in accordance with the same test, which is also accomplishable by the use of wood products.

In the draft system (2))there are no special references to wood products, apart from those which follow from their combustibility.

The work with NKB goes on - through the Sub-committee for fire protection - over the whole spectrum a) - e) above but has so far not resulted in complete drafts.

Further efforts within NKB are aimed at establishing a framework for the harmonization of building regulations, such as the following:

Level 1: Overall political goals for building
level 2: Basic definitions and principles (activities/measures)
level 3: Quantified, legally binding requirements
level 4: Methods of verification to control that requirements are being met
level 5: Examples of technical solutions "deemed to satisfy" the requirements.

Such a structure of levels implies, with regard to functional requirements and reference to standards:

- principles for the interpretation of society's goals into specific quantified requirements will be explicitly stated (level 2) and, consequently, subject to discussion;

- the potential for stating requirements in functional terms increases as goals are defined and interpreted into the requirements, and the methods of verification are analysed

- the policy and priority tasks for preparing standards in context of harmonizing building regulations may be defined (levels 4 and 5, in some cases level 3);

- the type of application of the reference to standards principle is indicated by the level on which it is to be applied.

It may be added that the ECE Committee on Housing, Building and Planning recently decided to recommend that a pilot study be undertaken on the basis of this hierarchy of levels, to help the work on building regulations harmonization within the Committee.

Building Regulations concerned with Fire Safety and their Impact on the Use of Wood Products in the Federal Republic of Germany

P. Frech
Otto-Graf-Institut, University of Stuttgart

FIRE PROTECTION IN THE FIELD OF PREFABRICATED HOUSES

The building regulations of the Länder require that buildings have to be designed, placed and established taking into consideration the protection of life and health (MBO building regulations § 3).

The following diagram demonstrates the influence of the legislator in the field of fire safety on the fabrication and use of building materials, the planning and construction of buildings. In general § 19 of the MBO deals with fire protection. "Buildings have to be placed, established and maintained in such a way as to prevent the development and propagation of fire and to allow effective fire-fighting methods and the rescue of men and animals during a fire".

Diagram 1

Influence of the Legislator on the Planning and Construction of Buildings in the Field of Fire-Safety

building regulations of the Länder (MBO §§ 3, 19 and others)

implementing ordinances and administrative stipulations

DIN4 102 and directives (directives for the use of combustible materials in surface building) introduced to the authorities by decrees

special stipulations

regulations for
- garages
- meeting places
- stores
- school buildings
- fire protection in industrial buildings (in preparation)

fabrication and use of building materials (choice of materials) planning and construction of the building (combination of the materials and construction of the elements)

1/ This paper draws on material contained in a speech made by Mr. C. Meyer-Ottens, Institut für Baustoffkunde und Stahlbetonbau (Institute of Materials Research and Reinforced Concrete Structures), University of Braunschweig.

The building regulations of the Länder are enlarged and specified by implementing ordinances and administrative decrees, as well as DIN standards, directives and supplementary stipulations. For schools, meeting-places, stores, garages and industrial buildings, special regulations have been issued (regulations for school buildings, meeting-places, stores, garages, and directives for the fire protection in industrial buildings are in preparation) which fulfil the special requirements concerning fire protection in these buildings.

DIN standards and regulations have the function of laws if they are introduced in the different Länder by the authorities, that is to say by a decree of the minister concerned. Thus they enter by this act of state into the "generally adopted rules of architecture" (MBO § 3/3).

The following have been introduced by the authorities in the field of fire protection:

1. DIN 4102 - behaviour in fire of building materials and structures

   Part 2: definitions, requirements and tests on building materials, February 1970 edition

   Part 3: definitions, requirements and tests on special structures, February 1970 edition

   Part 4: introduction into the definitions, February 1970 edition

2. Supplements to DIN 4102 - behaviour in fire of building materials and structures, February 1970 edition

3. Regulations for the use of flammable materials in building construction, May 1970 edition.

The last-quoted regulation forms the basis for the manufacture of wood panel houses according to DIN 1052. They stipulate that as an exception to the requirement of the building regulations to use only fire-resistant materials (for instance MBO § 32/3), it is possible - if there are no objections concerning fire safety - to use even combustible materials for special structures.

By DIN 4102 it is possible to estimate the fire resistance of building materials and structures.

The revision of the standard is finished. It will be published in the next weeks and introduced by the authorities in the summer of 1977.

The new edition of the standard DIN 4102 is divided into 8 parts:

Part 1: building materials; definitions, requirements and tests

Part 2: building structures; definitions, requirements and tests

Part 3: fire walls and non-bearing external walls; definitions, requirements and tests

Part 4: compilation and application of classified building materials and structures, and special elements

Part 5: fire doors, fire doors in elevator shafts and fire-resistant vitrifications; definitions, requirements and tests

Part 6: ventilation ducts; definitions, requirements and tests (in preparation)

Part 7: roofs; definitions, requirements and tests

Part 8: tests in a small-scale fire test set-up (in preparation).

BEHAVIOUR IN FIRE OF BUILDING MATERIALS

The new DIN 4102 maintains the division of the building materials into different classes according to their behaviour in fire (see table 1).

Table 1

Classification of Building Materials According to DIN 4102

building materials class	definition adopted by the authorities	example
A	incombustible materials	
A 1		concrete, gravel, plaster, asbestos-cement slabs, plaster-cardboard slabs
B	combustible materials	
B 1	flame-resistant materials	wood wool-lightweight slabs, wood materials treated by fire-protecting agents
B 2	normally inflammable materials	wood and wood materials more than 2 mm thick
B 3	easily inflammable materials	wood and wood materials up to 2 mm thick, paper, cardboard, rushes

The materials listed in the standard and divided into different classes need not be tested. All other materials require a testing mark given by the Institut für Bautechnik. By a certain fire-protecting treatment timber and timber materials become fire-resistant and may be classified B1 and not B2.

Plaster as a building material, as well as concrete, gravel and asbestos-cement slabs belong to class A1 without a special proof. Because of the addition of organic materials (paperboard) plaster-paperboard slabs need a testing mark and are classified A2.

Easily inflammable materials (B3) can only be used in connection with normally inflammable, flame-resistant and incombustible materials (see plaster paperboard slabs and plywood boards).

BEHAVIOUR IN FIRE OF BUILDING STRUCTURES

Building structures comprise walls, ceilings, beams, girders, etc. Their function is to resist the propagation of fire in the rooms, stories and fire sections.

The old as well as the new version of DIN 4102 makes a difference between fire resistance classes F30 to F180 (table 2).

Table 2

Fire Resistance Classes for Building Structures

Fire resistance class	Duration of fire resistance in minutes	Definition adopted by the authorities
F 30	30 to 60	fire-retardant
F 60	60 to 90	
F 90	90 to 120	fire-resistant
F 120	120 to 180	
F 180	180 and more	very fire-resistant

In the present and future edition of the standard, building structures erected in accordance with their applications in practice must maintain their defined functions, for instance maintenance of the bearing capacity and shutting off of a room, when exposed in a furnace to the influence of heat during a duration of 30, 60, 90, 120 or 180 minutes. The new edition of the standard deals with joints, connections and fixing devices mounted in accordance with their applications in practice.

Part 2 of the new edition of the standard proposes for the fire resistance class F 60, for which no definition has been stipulated by the authorities, the definition "very fire-retardant", for F 120 "very fire-resistant" and for F 180 "extremely fire-resistant".

Part 3 of the new standard describes the requirements and the classification of fire walls and non-bearing external walls (railings).

Table 3

Combination of Building Materials and Building Structure Classes according to DIN 4102, page 2, 1970 edition

F-class DIN 4102	Admitted materials class	Restrictions type	Materials class
F30 - F60	A - B	-	-
F90 - F120	A - B	1 - staticly important parts	A
		2 - coherent element, which must not be destroyed during the reference period: a) ceilings: A-thickness 50 mm b) walls: A-thickness: optional	A
		3 - walls thickness 50 mm	A
F 180	A	-	-

TABLE 4: Fire Safety Requirements for Outside and Inside Walls

type of wall		number of stories 1		2		≥ 3
		normal case	exceptions d/	normal case	exceptions d/	
outside walls bearing and stiffening	fire resistance class of the wall	F 30	no requirements	F 30	no requirements	F 90
	requirements to the material	B 2	B 2	B 2	B 2	mainly A
	insulating materials on the room side	B 1	B 2	B 1	B 2	B 1
	insulating materials inside the wall c/	B 1	B 2	B 1	B 2	B 1
	insulating materials on the outside	B 1	B 2	B 1a/	B 2a/	B 1a/
	coatings on the room side	B 2	B 2	B 2	B 2	B 2
	coatings on the outside	B 2	B 2	B 2a/	B 2a/	B 1a/

type of wall			number of stories 1	2	≥ 3	
outside walls	not bearing (for instance projecting elements)	fire resistance class of the wall	no requirements	no requirements	no requirements	F 30 / W 30
		requirements to the material	B 2	B 2	A	B 2
		insulating materials on the room side	B 2	B 2	B 1	B 1
		insulating materials inside the wall c/	B 2	B 2	A	B 1
		insulating materials on the outside	B 2	B 2a/	B 1a/	B 1a/
		coatings on the room side	B 2	B 2	B 2	B 2
		coatings on the outside	B 2	B 2a/	B 1a/	B 1a/
	not bearing, ≥ 150 m in every story projecting ceilings F 90	fire resistance class of the wall	not settled		no requirements	
		requirements to the material			B 1/B 2	
		Insulating materials on the room side and inside the wall c/			B 1/B 2	
		coatings on the room side				
		insulating materials on the outside			B 1/B 2 a/	
		insulating materials on the outside			B 1a/	

TABLE 4: Fire Safety Requirements for Outside and Inside Walls (continued)

type of wall			number of stories 1 and 2: normal case	number of stories 1 and 2: exceptions d/	≥ 3
inside walls	partition walls between dwellings, walls between rooms endangered by fire	fire resistance class of the wall	F 90	F 30	F 90
		requirements for the material	mainly A	B 2	mainly A
		insulating materials on the room side and inside the wall c/	B 1	B 1	B 1
		coatings	B 2	B 2	B 2
	bearing and stiffening	fire resistance class of the wall	F 30	no requirements	F 90
		requirements for the material	B 2	B 2	mainly A
		insulating materials on the room side and inside the wall c/	B 1	B 2	B 1
		coatings	B 2	B 2	B 2
	not bearing	fire resistance class of the wall	no requirements	no requirements	no requirements
		requirements for the material	B 2	B 2	B 2
		insulating materials on the room side and inside the wall c/	B 2	B 2	B 1
		coatings	B 2	B 2	B 2

type of wall			number of stories 1: normal case	1: exceptions d/	2: normal case	2: exceptions d/	≥ 3
inside walls of rescue ways	walls to stair-cases	fire resistance class of the wall or fire walls	F 90	no requirements or F 30	F 90	F 30	fire walls
		requirements for the material	mainly A	B 2	mainly A	B 2	A
		insulating materials on the stair-case side	A/B 1	B 2/B 1	A/B 1	A/B 1	A
		insulating materials inside the wall c/	B 1	B 2/B 1	B 1	B 1	A
		insulating materials on the room side	B 1	B 2/B 1	B 1	B 1	B 1
		coatings on the stair-case side	A	A	A	A	A
		coatings on the room side	B 2	B 2	B 2	B 2	B 2
	walls to generally accessible corridors	fire resistance class of the wall	no requirements		no requirements		F 30
		requirements for the material	B 2		B 2		mainly A
		insulating materials on the corridor-side and inside the wall c/	B 2		B 2		B 1
		insulating materials on the room side	B 2		B 2		B 1
		coatings on the corridor side	B 2		B 2		B 1
		coatings on the room side	B 2		B 2		B 2

For footnotes - see following page

TABLE 5: Minimum requirements of the authorities for building elements

1	Building elements	Offices and dwellings			
		up to 2 stories	up to 5 stories	up to multi-story houses	multi-story buildings
	1	2	3	4	5
2	Bearing and stiffening walls and supports	F 30 (fh) exceptions class B 2	F 90 (fb)		
3	Non-bearing external walls	F 30 or class B 2 exceptions (W 30)	class A or F 30 exceptions (W 30 or W 60)		exceptions (W 90)
4	Partition walls and the like	F 90 exceptions	F 90		
5	Non-bearing partition walls	class B 2	class B 2	class B 2	class A (nbr) exceptions
6	Ceilings	F 30 exceptions	F 30 class A	F 90	
7	Habitable attic walls and ceilings of an inhabitable attic (if not line 2)	F 30 exceptions	F 30 class A		
8	Walls of staircases	F 90 exceptions	F 90 thickness of fire walls (fire walls)		
9	Stairs	class B 2	class A	F 90	
10	Enclosure of escape routes (if not line 2)	class B 2	F 30 and class A		

a/ Building materials which "burn by dripping down" according to the building material class required, are not admitted.

b/ Structure of the wall: fire resistant wall, insulating materials class B 2 or B 3 and a layer of 6 cm of mineral materials.

c/ Higher requirements according to DIN 4102, p. 4 or testing certificates for materials of the classes F 30, F 90, W 30, W 90.

d/ Exception (§ 33 chapt. 3 MBO) for free-standing detached or double houses up to 2 stories and 2 dwellings and for other buildings of a similar size, 5 m distance between the boundaries, 10 or 8 m distance between the buildings, hard roofing.
For wood houses with prefabricated panels - one— storey building with garret or two-storey buildings - F 30 in the ground floor and insulating materials B 1.

If according to official statistics house building is subdivided into buildings with one (one-family houses), two (two-family houses) and more dwellings (multiple family houses) we see that especially one-family houses but also two-family houses and one-family ribbon-houses are built with prefabricated elements. In 1974 82% of all one-family houses, 13% of the two-family houses and only 5% of multiple family houses were prefabricated. It is estimated that 90% of all prefabricated one-family houses are built with panels of solid timber in connection with plywood, asbestos-cement and plaster cardboard slabs.

Tables 3 and 4 in connection with table 5 show that in buildings with more than 3 stories, combustible materials in connection with "non-combustible materials" (tables 4 and 5) cannot be used for staticly important elements because of the requirement F 90.

Up to now, it has not been allowed to erect buildings of more than 3 stories with prefabricated wood panels although bearing walls and ceilings have reached a fire resistance of more than 90 minutes. The connection between material classes and fire resistance classes has not been cancelled by the new DIN 4102 but reformulated.

For buildings of two stories, for which fire-retardant building elements are required, it is possible to use combustible materials, that is to say construction with wood panels fulfils the fire-protecting requirements.

The new as well as the old standard requires F 90 for non-bearing and non-stiffening as well as bearing and stiffening walls between one-family ribbon-houses. In the future the requirements for non-bearing and non-stiffening walls will be facilitated in so far as it is admitted to summarize the different resistances in order to obtain F 90.

One starts from the realistic assumption that for instance wall I resists for 45 minutes to the fire attack and only after that time is wall II affected by the fire. This wall has also a fire resistance of 45 minutes, so that the whole resistance reaches at least 90 minutes. The addition of the separate resistances is not valid for bearing and stiffening walls where A-Materials have to be used.

Part 4 of the standard constitutes an essential facilitation for prefabricated construction. The "compilation and application of classified materials, structures and special elements" makes it possible to choose among a great number of wall and ceiling constructions for which no separate tests are required.

The draft version of part 4 of DIN 4102 will probably be published in 1977.

FIRE PROTECTION IN THE FIELD OF INDUSTRIAL BUILDINGS

The draft standard DIN 18230 is related to the fire protection of industrial buildings.

This standard is accompanied by calculation methods by which the necessary fire resistance of industrial buildings should be determined.

On the basis of standard and valuation factors, a fire resistance value according to DIN 4102 is determined as a calculated value from which the fire safety class of the building or the fire section can be derived.

The necessary fire resistance classes of the different elements can be determined on the basis of the fire safety classes.

Fire safety measures which may result from the location and a special use of the building according to the prescriptions and directives of the authorities are taken for granted by the standard.

As the discussion about this comprehensive draft standard has not been finished and especially as the calculation factors may still be modified, we shall not enter into particulars here.

Building Regulations concerned with Fire Safety and their Impact on the Use of Wood Products: the Situation in Austria

H. Bednar
University of Agriculture and Forestry, Austria

INTRODUCTION

In Austria, regulations concerning fire and the use of materials for building construction date back to 1782. The imperial patents of that time contained clauses prohibiting thatched roofs and rooms for permanent living in the attic (it took nearly 200 years to overcome this instruction). Wooden staircases were not allowed in certain cases, and wooden chimneys were not allowed at all, special clauses existed concerning floors in kitchens and only vaults were allowed above stables, barns had to be erected out of villages. In rural areas, the distance between two houses had to be more than (appr.) 2m. The practical influence of the last direction can still be seen in certain parishes, especially in the eastern parts of our country.

Since the second half of the last century, the regulations were divided into those entirely concerned with the prevention of fire and the organization of fire extinction and others concerned with building and building codes.

In Austria, relevant regulations can be found within the building codes and in certain standards.

BUILDING CODES

Building and building codes are regulated by the individual provinces of Austria. This means that there exists no official single building code for Austria as a whole.

The "Forschungsgesellschaft für Wohnen, Bauen und Planen", Vienna, has created a model building code (Musterbauordnung) with contributions from a number of experts from all building fields and a great number of working groups. This model building code was revised several times and recommendations were made to take it as a rule for creating or revising building codes in the legislative assemblies of the different provinces.

Since the model code was established, all provinces have introduced new or revised building codes, but none of them has incorporated the model building code to any great extent. In fact, there are at present 13 valid building codes in existence, nine for the different provinces and four for larger towns in Salzburg and Upper Austria.

In addition to these particular building codes, regulations in by-laws and other laws influence building construction and the use of materials. Among them are regulations concerning the construction of theatres, circuses, school buildings and kindergarten buildings.

The existing 13 Austrian building codes have points in common but an even greater number of differences, which are very evident, for example, when measures and sizes are compared. In addition to that, some terms often used can have a different meaning within the different building codes. Some building codes have certain clauses concerning structural details and the choice of materials for buildings in rural areas, single buildings, buildings in suburban areas, buildings with one or two storeys and limited height.

THE FOLLOWING IS AN OUTLINE OF RELEVANT CLAUSES OF THE BUILDING CODE OF VIENNA, WHICH IS ONLY ONE OUT OF 13 IN AUSTRIA

Outer walls

Outer walls have to belong to the F 90 fire resistance class (see part: Standards), if nothing else is laid down. The F 90 class is not required for one-or-two-storey buildings and buildings in the outer green belt, but only in those cases where the location of the building or of the rooms requires an F 90 version for the outer walls. If the latter buildings contain residential rooms, the outer walls have to be made of an F 30 construction at least.

The parts of the outer walls between windows of the same storey need not be in an F 90 version in the case where the window area and the area of the wall not covered by F 90 classification, comprises less than 50% of the total outer wall of the appropriate room and regulations in respect of fire protection are not opposed. In cases where facade components are suspended to the outer wall, these parts have to be either within the fire resistance class F 90 or they have to be composed of non-combustible building materials. The latter have to be attached to the outer walls in such a way as to prevent the development of a chimney effect across different storeys.

Interior walls

All interior walls with the exclusion of partition walls within a compartment, if nothing else is specified, have to belong to the F 90 fire resistance class. In cases where partition walls are a part of an escape route of a workshop, they have to belong at least to the F 30 fire resistance class.

Comparmentation walls

Compartmentation walls have to be in the F 90 version in the following cases :

1) for rooms, where large amounts of materials with high fire risks are produced, processed or stored, with a maximum distance between compartmentation walls of 20m.

2) for workshops, storerooms and for guest-rooms (commercial enterprises) and for stables with a maximum distance of 30m.

3) within attics, if the interior height is only 0.6m and the outer walls and the ceilings below are made with an F 90 version with no openings for smoke ducts, and if heating and ventilition pipes are in existence, the regulation distance is of 90m ; otherwise it is 30m.

4) for basements with a maximum distance of 60m.

The attic has to be separated from dwellings, social rooms and internal wash-houses including their access paths and from machine workshops by an F 90 compartment wall. Doors within compartment walls have to be of F 30 version and self-closing.

For annex-buildings or outhouses, with the exclusion of buildings used to store large amounts of material with high fire risk, the compartmentation clause may not apply.

Wooden walls as outer walls and load-bearing walls

Exceptions from the clauses concerning the fire behaviour of wood walls are permitted as outer walls :

1) for one-storey buildings

2) for buildings belonging to class 1 (height restrictions)

3) for buildings within the green belt.

Outer walls made of wood and load bearing wood walls within a building, containing social or residential rooms, are to be made of F 30 version at least.

With the exception of annex-buildings or outhouses, wood walls may not be installed in place of compartment walls.

Wooden walls are not permitted as outer walls and load bearing walls if the location of the building or the purpose of the rooms makes it necessary to have F 90 version.

Ceilings and floors

The ceilings of buildings have to be, unless otherwise specified, of a construction of the F 90 class.

Ceilings made of wood need not belong to the F 90 class; their lower surface, however, has to be designed in such a way that it ranks within the F 30 class.

Wood ceilings are not permitted :

1) above cellars

2) on the top storey of buildings belonging to building classes II - IV, if they have more than one main storey and in all storeys of buildings ranked within building classes V - VI

3) above bathrooms, lavatories, and above and under rooms in which high moisture occurs

4) above rooms, in which large amounts of self-igniting, easily combustible or difficult to extinguish substances are produced, processed or stored, if above those rooms are social rooms or the only access route to such rooms

5) above and under large fireplaces

6) above rooms where transformers are installed.

In the case of wood ceilings, the floor has to be separated by a layer at least 8 cm thick. For this layer no substances may be used which are health-hazardous, destructive to the load-bearing components of the ceiling, flammable or rot-causing.

The requirement of having an F 90 version might not be necessary for one-storey buildings, for buildings belonging to building class I and for buildings within the green belt, if the location of the building or the designated use of the rooms do not require such a construction. Within these buildings, the ceilings of social rooms have to be at least of an F 30 resistance.

Beyond that, ceilings in two-storey buildings between flats only need to be loadbearing.

The loadbearing construction of ceilings and walls for social rooms in the attic has to be separated from wooden roof trusses by constructional designs.

The ceiling of the top storey has to be constructed in such a way that it supports, to a sufficient extent, the collapsing wooden roof trusses and the brickwork in the case of fire.

The floor construction has to be made of an F 90 version and the top floor layer has to be composed of non-combustible building materials :

1) in attics, with the exception of the above-mentioned clause (one-storey buildings, building class I, green belt)

2) in the case of large fire places, under that fire place up to a distance of 60 cm on the side of the opening and on the side of the ash extraction

3) in rooms in which large amounts of self-igniting, easily combustible or hard to extinguish substances are produced, processed or stored

4) within transformer rooms.

Roofs

The roof cover has to be sufficiently resistant against the effect of flames. Apart from that, the roof cover has to be made of non-combustible materials. Combustible materials might be used for buildings in the green belt and outhouses, if there is no hindrance to fire protection.

Roof constructions are not to be supported by wood ceilings.

Stairs, corridors

The ceilings and essential passageways have to have at least the same fire protection as the adjacent ceilings. In buildings with more than two main storeys the floor layer of all connecting passages have to have at least a difficult-to-ignite rating.

The staircase towards the cellar and the attic has to be partitioned by F 90 walls and ceilings, plus F 30 doors.

Stairs within buildings of more than two storeys have to be of an F 90 version. In buildings with less than two main storeys, the stair construction has to be at least an F 30 version. In dwelling houses with not more than two main storeys, stair constructions made of wood (conifer wood) are allowed even without an F 30 lining.

Small homes, terrace houses and summer cottages

By law, the government of the district of Vienna can take action to enforce for the above-mentioned three groups, the clauses concerning roof trusses, ceilings, and the F 90 class compartmentation of rooms in the attic.

STANDARDS

The essential standard for the matter under discussion is B 3800 : "Behaviour of building materials and components in fire", divided into five parts, three of which have already been published.

ONORM B 3800, part 1: Behaviour of building materials and structures in fire; Building materials (in preparation)

ONORM B 3800, part 2: Behaviour of building materials and structures in fire; Definitions, requirements and tests of structures.

The behaviour of structures in fire is characterized by their fire resistance and other conditions. Structures with special requirements are dealt with in part 3 of ONORM B 3800.

A grouping into different fire resistance classes is made according to the achieved duration of fire resistance:

Fire resistance class	Duration of fire resistance (in minutes)
F 30	more than 30 - less than 60
F 60	more than 60 - less than 90
F 90	more than 90 - less than 180
F 180	more than 180

CONDITIONS FOR FIRE RESISTANCE CLASSES

Fire resistance classes F 30 and F 90

During the tests the following conditions have to be fulfilled:

- Specimens of loadbearing structures may not collapse during the test whilst exposed to fire for a minimum time of 30 or 60 minutes, while bearing the calculated load.

- For building elements entirely or mainly stressed through bending during the fire test, the rate of bending has to be within a certain limit (rate of bending per minute less than the quotient of the span squared/8000 times thickness - both in centimetres).

- Normally the test has to be run with the designed bearing load applied. Tests without that load are only allowed if a testing laboratory, authorized in Austria, can calculate the fire resistance performance with accuracy. This is done on the basis of the technical data of the building materials of which the element is composed and other necessary information on its behaviour in fire, resulting from its reduction in cross-section and other alterations.

Specimens of space-covering elements have to prevent the penetration of fire, smoke or gaseous decomposition of products during the test for a minimum period of 30 or 60 minutes respectively. On the face of the specimen opposite to the fire, no flammable gases may occur, which could burn by themselves when the source of ignition is taken away.

At the end of the fire test, specimens of walls have to have all over the surface a minimum residual thickness of 1 cm, otherwise they have to withstand a special pendulum test.

Fire resistance classes F 90 and F 180

All components of the structure have to be made of non-combustible building materials according to part 1 of ÖNORM B 3800 (in preparation).

Proof of fire resistance classes

Without fire resistance test

All components mentioned in part 4 of ÖNORM B 3800 can be used in the given fire resistance class without proof.

With fire resistance test

For building components not mentioned in part 4 the fire resistance class has to be found by trials. For evaluating the fire resistance class two trials have to be made. Where one of the two gives a negative result, two more trials have to be made, both of which have to be positive.

The minimum number of test specimens is two. If there are different versions of a component in existence, the tests have to be made with the least appropriate version.

The temperature rise within the test chamber has to be in accordance with a given standard temperature curve; minor deviations are permitted. The means by which the temperature within and around the test specimen has to be recorded are also stated in the standard.

For space covering walls, a pendulum test is applied after the fire test.

Pillars have to be extinguished immediately after the fire test with a defined water source and direction.

Test construction for protective layers and covers

There are details given for ceilings made of steel beams, reinforced concrete and wooden beams.

In the case of wooden beams the moisture content has to be within the range of 15-20% and no flame retardants may be applied.

Detailed arrangements for the fire test of protective layers and covers for columns (steel reinforced concrete and wood) are given.

Finally, details concerning the test certificate are explained: among others - description and drawing of the test specimen, precise details of the building materials, sizes, weight units per area and per volume, amount of combustible materials, moisture content, means of fastenings for the protective layers, age of test specimen at the time of testing.

Observations are made during the test period covering areas such as: heavy development of smoke, highly irritating substances, sudden ignition, description of the specimen after the test.

ONORM B 3800, part 3: Behaviour of building materials and components in fire, special components, definitions, requirements and tests.

This part of standard B 3800 deals with components for outer walls, fire-resisting glazings, roof coverings, chimneys and chimney closures.

Components for outer walls

Components for outer walls are special components through which the spread of fire on the outer surface of the wall is made more difficult.

Load bearing outer wall components have to be tested according to ONORM B 3800, part 2.

Fire resistance classes for the above mentioned wall components

Fire resistance class	Fire resistance (in minutes)
W 30	more than 30 - less than 60
W 60	more than 60 - less than 90
W 90	more than 90

Tests with specimen

Components for outer walls have to be tested with their fixing devices and joints.

The inner part of the specimen has to be tested with a fire test according to the standard temperature curve (B 3800, part 2), the outer surface has to be tested with a fire test according to a special temperature curve (the temperature after 10 min. not to exceed 659°C).

ONORM B 3800, part 4: Behaviour of building materials and components in fire. Components: Classification in respect of fire resistance

Building components mentioned in this part of standard B 3800 do not need to be tested for fire resistance.

Building components not mentioned in the present part have to be tested according to part 2 or 3 of ONORM B 3800.

When applying the present standard, the behaviour of the total construction with regard to the entire building in the case of fire has to be considered. In particular, extensions resulting in disarrangements have to be taken into consideration. Linings and coatings made of combustible materials generally do not influence the classification with respect to fire resistance, but in some cases they have to be approved by the local building authority, especially when security is possibly influenced (e.g. meeting halls for public events, escape routes, undivided outer walls).

The following classifications refer only to the versions laid down in this standard.

BUILDING COMPONENTS WITH A FIRE RESISTANCE OF F 30

Walls and wall parts

Details of brick walls, concrete walls, light concrete walls, gypsum board walls, walls made of steel framework and walls made of wood and/or wood-based boards are mentioned.

For walls made of wood and/or wood-based boards, consisting of a framework made of scantlings and a cladding on both sides, the dimensions are given in detail, the joints of the single element are described; thermoplastic glues may not be used for joints stressed by fire.

Pillars and stanchions

Brickwork, reinforced concrete, steel;

Made of wood, also glue-laminated structures, centrally loaded,

(1) without linings or coatings with the following dimensions

longest buckling length (in m)	2.50	3.00	4.00	6.00
smallest sectional dimension (in cm)	16	18	20	22

(2) with protective linings according to details in the standard.

Ceilings

Bricks or prefabricated brick components, reinforced concrete, prestressed concrete slabs, steel beams with suspended plaster ceiling or suspended protective lining.

Made of timber with minimum sections of the beams according to Table 3 (see annex)

(1) Ceilings made of wood scantlings with protected lower surface (according to a special clause) and with non-combustible top layer or with a filling consisting of non-combustible material (thickness 5 cm).

(2) Ceilings made of single (visible) wood scantlings with a top layer made of 5 cm thick boards or a top layer made of 4 cm thick particle board, minimum density 600 kg/m^3, with single tongue and groove or supported joints.

(3) Ceilings made of wood scantlings and with a top and lower surface made of wood or wood-based boards and a fire protective layer made of mineral wool, minimum density 50 kg/m^3, thickness 6 cm, supported by and fixed on a wire mesh.

(4) Layer for ceiling surface made of matchboards, 4 cm thick, with double tongue and groove, tightly closed, or particle board, thickness 3 cm, minimum density 600 kg/m^3, joint support; or wood wool slabs with mineral binders, 5 cm thick, joints have to be smoothened, the bottom surface of the wood wool slabs has to be closed with mineral binders.

Girders, supports, crossbars

Reinforced concrete, prestressed concrete, steel;

Wood, also glue laminated

(1) without plastering, without linings, rectangular section, minimum cross-section 450 cm^2, minimum dimensions 12 cm (width) and 20 cm (height)

(2) with protective linings according to a special clause.

Stairs within an F 90 stairway

Steps made of concrete, situated on a support made of concrete or supported on a steel construction.

Stairs made of wood, steel or stone, bottom surface and side supports protected.

Stairs construction made of oak wood, side support and steps, all components with a minimum thickness of 5 cm.

BUILDING COMPONENTS WITH A FIRE RESISTANCE F 60

Walls and wall components

Brick walls, concrete walls, reinforced concrete walls, gypsumboard elements, steel framework.

Wood and/or wood-based boards, consisting of a framework made of solid wood and claddings on both sides, the outer layer of the cladding has to be made of non-combustible material.

Pillars and stanchions

Brickwork, reinforced concrete steel;

Wood, also glue-laminated, centrally loaded

(1) without protective linings, dimensions

longest buckling length (in m)	2.50	3.00	4.00	6.00
smallest sectional dimension (cm)	27	28	30	35

(2) with protective linings according to a special clause.

Ceilings

Brick or prefabricated brick-components, reinforced or prestressed concrete construction, reinforced concrete slabs, vaults made of concrete, brick or concrete blocks, steel profiles.

Wood, the necessary cross-sections of the beams are given in Table 3 (see annex); for the protective coverings details are given in Table 1, 2 and/ or 4.

(1) Ceilings made of wood scantlings with covered lower surface according to a special clause. The top layer must be non-combustible or a filling with a non-combustible material has to be given, thickness 7 cm.

(2) Ceilings made of single (visible) wood scantlings with a top layer made of 7 cm thick boards with double tongue and groove, tightly closed.

Girders, supports and cross bars

Reinforced concrete, prestressed concrete, steel;

Wood, also glue-laminated

(1) without protective lining, with a minimum rectangular section of 1200 cm^2, minimum dimension 20 cm (width) and 32 cm (height)

(2) with protective layer according to a special clause.

Stairs within an F 90 staircase

Wood not allowed.

BUILDING COMPONENTS WITH A FIRE RESISTANCE OF F 90 OR F 180

Combustible materials may not be used in any part within components of the two fire resistance classes.

TABLE 1 Combustion rate of wood and wood-based boards

Material	Rate of combusion (mm/min)
Timber for constructional purposes according to ÖNORM B 4100 part 2	
Oak	0.50
Spruce, including glue-laminated	0.65
Pine	0.75
Particle board according to ÖNORM B 3002	
Density 600 - 700 kg/m^3	0.80
Density more than 700 kg/m^3	0.70
Plywood according to ÖNORM B 3008	
Beech	0.80
Spruce	0.80
Fibreboard	
Hardboard	0.65
Insulating board	2.00

TABLE 2 Fire resistance duration of fire protective layers (without plaster)

Material	Thickness (cm)	Density (kg/m^3)	Duration of fire resistance (in minutes)
Mineral wool mats	5	50	15
	10	30	15
Woodwool slabs with mineral binders	3.5	400	20
	5	400	25
Boards made of mineral fibres	5	300	30
	2 x 1.9 a/	400	35
	2 x 1.3 b/	500	40
Gypsum boards with kraftliner	1.25	840	15
	1.5	840	18

a/ Construction with space between

b/ Construction with 4.5 cm distance between boards

TABLE 3 Minimum cross-sectional dimensions for framework of walls and ceiling beams made of wood or wood-based materials

Building component	Fire resistance class	Minimum dimension of scantlings			Enlargement of cross-sectional dimensions based on load-bearing calculations in case of a lining or protective layer with a fire resistance of		
		section (cm^2)	width (cm)	height (cm)	0 min	15 min	30 min
Walls							
non-load bearing	F 30	25	4	6	0	0	0
	F 60	40	4.5	9.5	0	0	0
load bearing	F 30	96	8	10	1.5	0	0
	F 60	250	14	16	3	1.5	0
Ceilings							
	F 30	140	10	12	1.5	0	0
	F 60	320	16	18	5	2.5	0

TABLE 4 Minimum thicknesses of linings (plasters, coverings) for wood and steel building components to come up to fire resistance class F 30, F 60 and F 90 according to ÖNORM B 3800, part 2

Kind of building material for the lining	Minimum thickness (in cm) for the fire resistance classes		
	F 30	F 60	F 90 (for steel components only)
Plasters			
plaster made of cement	2.0	4.0	6.0
plaster made of lime/cement, lime/gypsum, gypsum/sand	1.5	3.0	4.5
special cement or gypsum mortars or plasters (maximum density 1500 kg/m^3)	1.5	2.5	3.5
sprayed asbestos layer with mineral binders	2.0	2.5	3.0
Boards and slabs			
woodwool slabs with mineral binders according to ÖNORM B 3465 (with plaster and cement layer)	2.5	5.0	..
light concrete slabs (density less than 1600 kg/m^3)	4.0	4.5	5.0
gypsum boards	3.0	3.0	5.0
concrete slabs with light inorganic additives (density less than 1300 kg/m^3)	3.0	3.5	4.0
gypsum boards with kraftliner coat according to ÖNORM B 3410, joints bonded and smoothened			
version GKB	1.8		
version GKF	1.25	2 x 1.25	3 x 1.50
asbestos-cement slabs with light inorganic additives according to ÖNORM B 3425			
for wood components	1.5	3.0	..
for steel components	1.0	2.0	3.0

Influence des Préscriptions de Police du Feu sur l'Emploi du Bois et des Matériaux Dérivés du Bois dans la Construction en Suisse

B. Gfeller

RESUME

Les mesures de protection contre l'incendie mises en oeuvre dans la construction pour sauver les vies humaines et les biens matériels en cas d'incendie s'occupent notamment de l'influence que peuvent avoir, sur le phénomène feu, les matériaux et éléments de construction utilisés dans un bâtiment. Dans ce domaine, une des tâches essentielles consiste à étudier et apprécier le comportement de ces matériaux. Les résultats de ces recherches se retrouvent dans les dispositions de police du feu qui sont édictées par les autorités cantonales compétentes.

INTRODUCTION

Puisque l'arrêté des prescriptions de police du feu est du ressort des cantons, il existe actuellement dans ce domaine autant de normes juridiques qu'il y a d'Etats dans la Confédération. A cela s'ajoute le fait que, d'un canton à l'autre, les formes d'organisation sont différentes, surtout en ce qui concerne les relations de subordination des instances chargées de la protection contre l'incendie, si bien que l'ensemble des problèmes relatifs à la protection contre l'incendie se révèle être d'une grande complexité.

Depuis quelque temps néanmoins, et grâce à l'activité déployée par l' "Association des établissements cantonaux d'assurance contre l'incendie" (AECA) on se dirige vers une certaine unification dans le contenu des prescriptions de police du feu. Cette association est un groupement du droit privé, qui rassemble les établissements cantonaux d'assurance contre l'incendie qui ont, eux, un statut de droit public. De plus, avec le temps, les instances de protection contre l'incendie des cantons ne disposant pas d'établissements d'assurance publics ont aussi adhéré à l'association. Les publications de celle-ci, en particulier ses "Directives pour les prescriptions sur la police du feu", trouvent de plus en plus d'écho dans les dispositions cantonales, où elles sont parfois reprises mot pour mot.

Une autre tâche très importante de l'association consiste à établir à l'usage de ses membres des recommandations de caractère non contraignant sur les matériaux qu'on peut admettre dans la construction, étant donné leur comportement au feu. Rédigées par la Commission technique de l'association, ces recommandations suscitent le plus vif intérêt parmi les membres. Et nombreux sont les cantons qui tiennent compte des avis émis par l'association pour admettre un matériau ou une partie de construction à des fins bien précises. C'est pourquoi, dans ce qui suit, notre attention se portera principalement sur les "Directives pour les prescriptions sur la police du feu".

1/ Ce document est reproduit avec l'autorisation de AG Verlag Hich- und Tiefbau

DEFINITIONS TECHNIQUES, EN FONCTION DE LA PROTECTION CONTRE L'INCENDIE, DES MATERIAUX, ELEMENTS ET PARTIES DE CONSTRUCTION

Dans ses "Dispositions générales", l'Association des établissements cantonaux d'assurance contre l'incendie distingue entre la réaction du feu des matériaux et éléments de construction d'une part, et la résistance au feu des parties de construction d'autre part.

Les matériaux de construction sont des substances utilisées pour la fabrication d'éléments et de parties de construction.

Les éléments de construction sont faits de matériaux de construction et servent à la fabrication de parties de construction.

Les parties de construction sont les parties constitutives d'un ouvrage et sont faites de matériaux ou d'éléments de construction.

On entend par réaction au feu d'un matériau ou d'un élément de construction la réaction de ce matériau ou de cet élément aux températures élevées qui se produisent en cas d'incendie.

Enfin, la résistance au feu d'une partie de construction se définit comme la propriété assurant le maintien, sous l'action d'un incendie, des fonctions incombant à la partie de construction dans l'ouvrage.

Pour la réaction au feu, on distingue entre matériaux de construction combustibles et matériaux de construction incombustibles. Ces deux groupes principaux se subdivisent à leur tour en classes, dont on trouve les dénominations au tableau 1.

On n'a pas représenté dans ce tableau l'intensité des fumées, parce que c'est un facteur négligeable en ce qui concerne le bois. Il pourrait toutefois gagner en importance étant donné les efforts accrus entrepris pour ignifuger les matériaux à base de bois.

De même, la corrosivité et la toxicité des gaz qui se dégagent pendant un incendie entrent dans l'appréciation d'un matériau de construction quant à sa réaction au feu. Etant donné que cet aspect n'est pas non plus très important pour le bois ou les matériaux à base de bois, nous laisserons également de côté ces problèmes dans ce qui suit.

Comme on peut le voir au tableau 2, les parties de construction sont rangées dans des classes de résistance au feu, en fonction de critères techniques relevant de la protection contre l'incendie. La durée de résistance au feu mesure le temps durant lequel une partie de construction est capable d'empêcher l'extension du feu, tout en pouvant remplir les fonctions qui lui sont imparties dans l'ouvrage. Par "parties spéciales", on entend essentiellement des portes ou des fermetures de segments de feu.

D'après les considérations contenues dans les dispositions générales des "Directives pour les prescriptions sur la police du feu", on a rangé dans les diverses classes les matériaux, les éléments et les parties de construction en se fondant sur des méthodes de contrôle reconnues. Pourtant on ne trouve de description de ces méthodes que concernant la réaction au

Tableau 1: Classification, méthodes d'examen et exigences concernant la réaction au feu des matériaux
(Tiré du projet de directives de l'Association des établissements cantonaux d'assurance contre l'incendie sur l'"Examen des matériaux et éléments de construction", et des "Feuilles du SPI" BA/1/2370 et SW/2/2205-6) (Service de prévention d'incendie pour l'industrie et l'artisanat, Zurich)

Classe	Dénomination	Exemples	Méthodes d'examen		Caractéristiques de classification	Exigences	
			Test de base A	Test supplémentaire B		A	B
I	S'enflammant très facilement et se consumant très rapidement	Pas admis comme matériau de construction					
II	S'enflammant facilement et se consumant rapidement		Inflammation des bords 15s., sous un angle de 45°		Pointe de flamme au bord supérieur avant	5s.	
III	Facilement combustible	Copeaux de bois liège de construction	id. cl. II	Inflammation de la surface, 45s., sous un angle de 45°	Pointe de flamme au bord supérieur avant	20s.	40s.
IV	Moyennement combustible	Bois de sapin panneaux de bois aggloméré et de fibres	id. cl. II	id. cl. III	Pointe de flamme au bord supérieur après L'échantillon s'éteint après	20s. 20s.	40s. 50s.
V	Difficilement combustible	Bois de chêne panneaux de particules ignifugés	id. cl. II	id. cl. III	L'échantillon s'éteint avant	20s.	50s.
VIq	Quasi incombustible	panneaux de particules liés au ciment	Selon les dispositions complétant DIN 4102				
VI	Incombustible	Métaux, etc.	notamment pouvoir calorifique inférieur à			1000kcal/kg	

feu, et cela seulement en esquisse. (Voir tableau 1). Il est vrai que ces méthodes sont appliquées depuis un certain temps déjà au LFEM. Et pour la résistance au feu, ce sont les résultats actuellement acquis au LFEM, selon un procédé bien introduit sur le plan international, qui sont retenus. Le tableau 2 représente les exigences auxquelles doivent satisfaire les parties de construction, et notamment les parties spéciales, pour entrer dans l'une ou l'autre classe. Il y apparaît que les parties de construction soumises au test d'après la courbe unitaire de température (DIN 4102) sont appréciées essentiellement d'après les critères suivants :

- Etanchéité au passage des flammes et des gaz
- Perte de force portante pour les parties portantes
- Isolation thermique.

Pendant le test, chacun de ces trois critères est mesuré et observé en permanence. On ne tolère en outre pendant l'examen aucune inflammation du côté tourné vers le foyer de l'incendie. Cette exigence n'est levée que pour les constructions lamellées collées (charpente Hetzer), certains revêtements de plafonds et pour les classes de parties spéciales T30 et T60.

CLASSIFICATION, EN FONCTION DE LA PROTECTION CONTRE L'INCENDIE, DES BOIS ET DES MATERIAUX DERIVES DU BOIS

Le tableau 3 regroupe les résultats de quelques recherches menées au LFEM sur la réaction au feu du bois massif. Des six essences de bois testées, cinq sont à l'état sec, moyennement combustibles. Seul l'épicéa est facilement combustible. Avec une teneur en eau de 8 à 9 %, telle qu'on la constate en

Tableau 2 : Classification et exigences concernant la résistance au feu des parties de construction
(Tiré des dispositions générales des "Directives pour les prescriptions sur la police du feu" de l'Association des établissements cantonaux d'assurance contre l'incendie)

Classe	Durée de résistance au feu (min.)	Dénomination	Exemple	Exigences
Parties de construction				
F 30	30	Entravant la propagation du feu	cloison de séparation entre un bureau et une issue Tabliers à feu	— Aucune inflammation du côté tourné vers le feu: - Aucun passage des flammes et des gaz inflammables: - Elévation de température de l'autre face: en moyenne ≤ 140°C, au maximum 180°C: - Pour des revêtements de constructions portantes en acier, élévation maximale de température 400°C (matériaux combustibles admis): - Aucune perte de force portante
F 60	60	id. F 30	—	— id. F 30, mais pendant 60 minutes
F 90	90	résistant en feu	Mur mitoyen, cloison entre un bureau et un corridor dans un immeuble-tour	— id. F 30, mais pendant 90 minutes et pour des parties de construction portantes et de fermetures d'espaces: - Matériaux et parties de construction de la classe VI: Solidité résiduelle plus élevée que la pression de l'eau à éteindre l'incendie
F 120	120		—	— id. F 90, mais pendant 120 minutes
F 180	180	hautement résistant au feu	—	— id. F 90, mais pendant 180 minutes
F 240	240		Mur pare-feu	— id. F 90, mais pendant 240 minutes
Parties spéciales				
T 30	30	Entravant la propagation du feu	Porte d'entrée d'un logement	— id. F 30, mais: - Inflammation tolérée du côté tourné vers le feu: - Température maximale de l'autre côté: jusqu'à 200°C (bords)
T 60	60	id. F 30	—	— id. T 30, mais pendant 60 minutes
T 90	90	résistant au feu	—	— id. T 30, mais pendant 90 minutes: - Matériaux et éléments de la classe VI

moyenne dans les locaux d'habitation, tous les bois de feuillus qui ont été examinés sont difficilement combustibles. Avec une teneur en eau de 13 à 15 %, comme elle peut se produire exceptionnellement dans certains locaux, toutes les essences de bois examinées sont difficilement combustibles. Mais, outre l'humidité du bois, le poids spécifique de celui-ci est aussi d'une grande importance. Ainsi les essences de bois légères témoignent d'une réaction au feu en général moins bonne que les essences lourdes.

Tous les matériaux à base de bois, sous forme de panneaux - à l'exception de certains panneaux de particules ignifugés (classe V) ou liés au ciment (classe VIq) - sont moyennement combustibles (classe IV). Etant donné que la réaction au feu de ces matériaux dépend dans une forte mesure de la proportion de bois qu'ils contiennent, il nous paraît indiqué ici de jeter un bref coup d'oeil sur le mécanisme de combustion du bois. Sans tenir compte du facteur temps, ce mécanisme peut se décompter pour l'essentiel selon les étapes suivantes :

1. Jusqu'à 150°C environ, on constate que se produit la déshydratation (y compris l'eau de cristallisation);

2. A partir de 150°C et jusqu'à 275°C environ, du CO_2 incombustible en quantités importantes se dégage. Parallèlement à ce processus, on remarque un changement de couleur, parfois déjà de la carbonisation;

3. A partir de 300°C environ, des hydrogènes de carbure fortement combustibles sont ensuite libérés, qui s'enflamment spontanément et qui - étant donné leur pouvoir calorifique élevé - entraînent une rapide extension des flammes. Ce processus atteint son point culminant vers 450°C;

4. Suivant cette phase décisive pour le phénomène global du feu, une autre phase a lieu, dont l'activité est notablement moindre et qui est caractérisée par la fiscion d'hydrogène et d'oxyde de carbone, qui ne donnent plus au feu qu'une modeste impulsion.

Il apparaît comme très important, en particulier pour l'étude de la résistance au feu des parties de construction, que la température critique pour le bois et les matériaux dérivés du bois se situe aux alentours de 300°C. En effet, dans ce test, la température monte durant les trente premières minutes à plus de 800°C déjà. Il est donc inévitable durant la première demi-heure que le côté exposé au feu des parties de construction en bois et matériaux à base de bois s'enflamme; on ne peut donc à priori ranger de tels matériaux dans la classe de résistance la plus faible au feu. Sont considérés comme faisant exception, nous l'avons déjà dit, certaines constructions et certaines parties spéciales bien précises.

IMPORTANCE DE CETTE CLASSIFICATION POUR LA MISE EN OEUVRE

Comme, sur la base des tests pratiqués à l'heure actuelle, le bois et les matériaux tirés du bois doivent être considérés comme combustibles, leur domaine d'utilisation en tant que matériaux de construction, figure, dans les directives à paraître prochainement de l'Association des établissements cantonaux d'assurance contre l'incendie, sous la rubrique "Emploi de matériaux de construction combustibles dans les bâtiments". Mais ces directives ne sont encore qu'à l'état de projet; c'est pourquoi nous ne pouvons les apprécier ici que d'une manière très générale. Leur principe consiste à distinguer les bâtiments selon leur nombre d'étages et leur genre d'affectation. Plus le nombre des étages devient élevé, plus les prescriptions se font sévères. Ainsi, on ne peut utiliser pour les parties portantes d'une construction des matériaux de construction combustibles que jusqu'au deuxième étage. Mais déjà à l'intérieur de tels bâtiments, on ne doit employer que des matériaux incombustibles le long des issues. La même exigence s'applique aux bâtiments dont le genre d'affectation implique de grands rassemblements de personnes (cinéma, théâtre, etc.). Dans les immeubles-tours (huit étages et plus, ou 22 m de haut), seuls des matériaux incombustibles peuvent être employés. La question reste encore ouverte de savoir dans quelle mesure ces nouvelles prescriptions influeront sur la pratique aujourd'hui courante, puisque le détail de ces prescriptions n'est pas encore connu.

En ce qui concerne la résistance au feu, on constate que le bois et les matériaux dérivés du bois - sous la forme de parties de construction - se trouvent négativement affectés par le fait qu'on ne peut les ranger à priori dans la classe de résistance la plus faible au feu. C'est là une constatation qui se répète sans cesse, s'agissant en particulier des cloisons de séparation pour lesquels les matériaux dérivés du bois sont particulièrement bien appropriés. En République fédérale d'Allemagne, où l'on effectue tout comme en Suisse les tests de résistance au feu, on n'avance pas cette même exigence selon laquelle le côté tourné vers le feu ne doit pas être enflammé. Grâce à cela, un grand nombre de construction des classes F30 et F60 sont admises comme cloisons de séparation, qui sont exclues en Suisse. Le fait même d'avoir recours à la réaction au feu comme critère de classification des résistances au feu constitue en fait un préjudice porté au bois et aux matériaux dérivés du bois.

Pour apprécier le danger d'incendie présenté par un matériau ou une partie de construction, il faut en principe partir des deux aspects suivants:

1. La perte, en cas d'incendie, des propriétés fondamentales justifiant qu'on ait utilisé cette partie de construction; ainsi par exemple, la perte de la force portante, de la forme, et, par conséquent, de la fonction de fermeture d'un espace, etc.

2. La contribution que le matériau de construction dont est faite la partie de construction peut apporter à la propagation de l'incendie.

La perte des propriétés est une conséquence de l'élévation de la température, qui entraîne des altérations physico-chimiques. L'élévation de la température d'un matériau de construction est fonction de son indice de pénétration thermique. Plus cet indice est élevé, plus le matériau s'échauffe rapidement. Or il apparaît au tableau 4 que, comparés à d'autres matériaux de construction, le bois et les matériaux dérivés du bois se caractérisent par des indices de pénétration thermique considérablement plus bas, ce qui signifie qu'en cas de feu, ils s'échauffent plus lentement, atteignant plus tard la température critique. Ce fait, conjugué à un coefficient de dilatation thermique deux à trois fois plus bas (tableau 4) signifie que le bois, pendant la première phase de l'incendie, c'est-à-dire au moment où devraient intervenir les mesures de secours, se comporte mieux que ne le révèlent les résultats obtenus aux examens de matériaux. Certes, possédant une puissance calorifique de quelque 4000 kcal./kg - libérée en partie à des températures assez élevées ($> 300^{o}C$) - les bois et les matériaux à base de bois fournissent une impulsion à l'incendie. Mais cela n'est vrai que dans certaines conditions données (teneur en humidité, rapport surface/volume, poids spécifique, grandeur de la pièce, etc.). On sait par exemple que certaines poutres lamellées, à des dimensions adéquates, peuvent résister très longtemps à l'agression du feu sans stimuler la propagation des flammes. Des recherches personnelles nous ont en outre démontré que, de même, des panneaux de particules d'une certaine épaisseur ne libèrent qu'un fragment de leur puissance calorifique théorique. Cette observation, dont l'explication se trouve dans l'effet de protection assuré par la couche de charbon qui se forme, s'est trouvée confirmée par le fait que ces matériaux, une fois éloignés de la source d'inflammation, ne continuent pas à brûler d'eux-mêmes. En procédant, durant la fabrication du panneau de particules, à une ignifugation supplémentaire, on arrivera à un abaissement figurant une moindre libération du pouvoir calorifique.

Compte tenu des considérations que nous venons d'exposer, on ne saurait donc se satisfaire d'un test de réaction au feu se fondant sur une inflammation brève et sur un pouvoir calorifique purement théorique, obtenu par combustion forcée seulement, et à partir duquel on apprécierait d'un point de vue technique de protection contre l'incendie la valeur du bois et des matériaux tirés du bois.

Tableau 3 - Indices du danger d'incendie présenté par six essences de bois après entreposage dans diverses humidités relatives de l'air

	100°C		20°C + 35% HR		20°C + 70% HR		18°C + 90		
	Teneur en eau %	Classe	Teneur en eau %	Classe	Teneur en eau %	Classe	Teneur en eau %	Classe	Poids spécifique moyen à l'état sec kg/m3
Bouleau	0	IV	8,1	V	14,3	V	18,6	V	608
Hêtre	0	IV	8,1	V	12,9	V	18,5	V	645
Chêne	0	IV	9,1	V	13,9	V	18,2	V	582
Frêne	0	IV	8,7	V	13,6	V	18,4	V	682
Epicéa	0	III	9,2	IV	15,2	V	19,6	V	366
Sapin	0	IV	9,1	IV	13,8	V	19,6	V	362

Tableau 4 - Indices de pénétration thermique et coefficients de dilatation thermique de divers matériaux de construction

Matériau	Indice de pénétration thermique (Kcal/m2/H 1/2°C)	Coefficient de dilatation thermique μ m/m°C
Acier	260	12
Béton	28	13
Béton armé	31	15
Maçonnerie	17	5
Bois massif (15 % d'humidité)		
Chêne	9,2	3-6
Epicéa	6,1	3-6
Panneau de fibres de bois		
Dur	6	-
Mou	2	-
Panneau de particules	6	Négligeable

CONCLUSIONS

Les "Directives pour les prescriptions sur la police du feu", dues à l'Association des établissements cantonaux d'assurance contre l'incendie, suscitent aujourd'hui un très grand intérêt, et les cantons y ont recours comme ligne directrice de leur activité légiférante en ce domaine. Elles revêtent donc une immense importance pour l'admission ou le refus de tel ou tel matériau et de telle ou telle partie de construction étant donné leur valeur technique de protection contre l'incendie. Or les prescriptions aujourd'hui en vigueur portent un réel préjudice aux matériaux de construction en bois et dérivés du bois, ainsi qu'aux parties de construction faites en bois ou en matériaux dérivés du bois. Il est de fait qu'elles en limitent l'emploi dans des domaines d'utilisation fondamentaux. Il en est ainsi non seulement dans l'intérêt d'une sécurité plus grande, mais aussi en raison de procédés d'examen dont on peut douter qu'ils "collent" à la réalité pratique. D'une part, on y fait appel à des caractéristiques de classification (inflammabilité, pouvoir calorifique lors de l'examen de réaction au feu, inflammabilité du côté de l'échantillon tourné vers le feu lors de l'examen de la résistance au feu) qui sont à priori négatives pour le bois et les matériaux dérivés du bois; d'autre part, on n'y tient pas compte des facteurs favorables, dont il a été prouvé qu'ils ont une influence positive en cas d'incendie.

On comprend aisément qu'on ne pourra jamais traduire toute la complexité du phénomène feu par un seul procédé d'examen, qui se doit d'être aussi "maniable" que possible. Mais il faut aussi que soit bien claire la nécessité de ne pas causer de tort, dans des proportions qui vont à l'encontre de ce que révèle un examen minutieux des faits, à des matériaux d'une importance capitale, sur la base de tels procédés et des résultats auxquels ils mènent.

Fire Regulations and Use of Wood-based Products in Finland

Tarmo Vilppula
Woodinformation, Finland

1 FIRE REGULATIONS AND STRUCTURES

New fire regulations came into force in Finland on 1.7.1976. Though they made no radical change in the situation regarding wood-based products, they broke new ground in form and procedures. Positive features in them are their attention to a performance concept and their alignment with the rules of other Scandinavian countries.

With reference to wood-based products these regulations may be examined by grouping the areas of application as follows: low-[1] and high-rise dwelling houses; buildings other than residential.

1.1 Low-rise dwelling houses

Ease of escape when danger threatens is a natural feature of the low-rise apartment. Personal safety is therefore no great problem.

Low-rise houses are built in Finland for one, two or more families with 1, 1½ or 2 storeys. In the last case the same family occupies both storeys. In 1976 some 26,000 low-rise dwellings were built in Finland, three-quarters of them wood-framed.

It may be regarded as the main rule in low-rise housing that wood-based products are used without restriction. One exception is porous fibreboard, whose use as interior facing is limited to one-storey houses.

Low-rise housing may be built wood-framed if the gross floor area does not exceed 400 sq.m. for a one-storey structure, and 200 sq.m. for two storeys, unless the building is fire-compartmented.

Compartmentation can be made either dwellingwise with building components of the B 30 class[2], at least, or in compartments up to 400 sq.m. with building components of at least B 90 or A 60 class[3]. When compartmented in this way, the gross floor area may be up to 2400 sq.m. for a one-storey building.

---

[1] These include terrace houses in accordance with Finnish practice.

[2] Components which may contain combustible building materials and have a fire resistance of at least 30 minutes.

[3] A component of incombustible material, with fire resistance of at least 60 minutes.

Two-storey buildings may be divided into compartments of separate dwellings or compartments up to 200 sq.m. by using separating structures similar to those used in one-storey buildings. Maximum permitted gross floor area is then 1600 sq.m.

Several buildings of the type mentioned may be constructed together if separated by fire walls.

If separate residential buildings for one or two families are erected less than 8 metres apart, the aforementioned gross floor areas are still permissible. This is on condition, however, that the walls of neighbouring buildings at less than the above-mentioned distance are of at least class B 30 structure, that the distance between buildings is at least 4 metres, and that each of the above-mentioned adjacent walls contains only separate windows of up to 0.2 sq.m., total window area up to 1.0 sq.m.

1.2 High-rise dwelling houses

1.2.1 Outer walls

In outer walls combustible building materials may be used only in exceptional cases. These are:

- Small components may be faced with combustible materials. In Finland some advantage is taken of this for the adornment of 3- and 4-storey houses.

- In buildings of up to 8 storeys recessed portions of outer walls (i.e. interior walls of balconies) may be faced with combustible materials under certain conditions. In that case the wall framework may also be of combustible materials. Wooden recessed balcony walls of this nature are very common in Finland.

- In buildings of up to 8 storeys the outer wall, if not load-bearing, may be framed with combustible materials provided the outer facing is incombustible and the frame is protected with interior facing of at least B 10 class, also that the ventilation gap is lined with material whose surfaces are of ignitability class 1 and fire spread class 1. Such structures are very rare in Finland due to the old fire regulations, which allowed them only by special permission. Now that the regulations have changed, the matter becomes relevant. In Sweden and Norway such an outer wall structure is used for at least 4/5 of all new high-rise apartment houses up to 8 storeys.

1.2.2 Building components for compartmentation

Non-loadbearing components must be of at least B 60 class in buildings of up to 8 storeys. In principle, therefore, these components may be of combustible materials such as wood.

1.2.3 Inner walls of compartments

For the inner walls of a compartment combustible materials may be used in buildings of up to 8 storeys provided the surface layers of these materials fulfil certain requirements.

The requirement mentioned, ignitability class 2[1] is fulfilled by all wood products except porous fibreboard.

In Finland the inner walls of a compartment, i.e. light partition walls, are generally wood-framed: 34 x 70 c/c 60.

1.2.4 Surfaces of indoor ceilings and walls

In high-rise apartment houses, these may be of combustible materials which meet class 2 ignitability requirements, i.e. wood products.

In Finland wood products are used in surface layers mainly for light partition walls and floors (parquet).

1.3 Buildings other than residential

The whole of the foregoing has referred to normal residential buildings either low- or high-rise housing, which are used for ordinary living purposes.

In fire regulations accommodation is grouped according to its function and the extent to which human safety is endangered by an outbreak of fire.

Apart from residential housing these groups include: public accomodation such as hospitals, children's homes and prisons; day care centres for children; places of congregation such as shops, exhibition halls, restaurants and working premises.

For these premises there are fire regulations which differ in some respects from the above and which need not be touched on here. It should be understood that they often impose greater restrictions on the use of wood-based products than regulations for residential housing.

2 FIRE REGULATIONS AND THE SAFETY QUESTION

With reference to the use of wood-based products Finland's fire regulations may on the whole be regarded as suitable and expedient.

The matter may, of course, be examined from another viewpoint: how have fire regulations in practice met the needs of human security? As stated, Finland's new regulations were adopted on 1.7.1976, so that their effect cannot yet be discussed. But let us look at what occurred earlier.

In 1970 Finland's housing as recorded in the census was divided thus: of the 1.5 million dwellings in use, at least 60% or 925,000 were in wooden buildings and 575,000 in buildings of stone or brick. In round figures, therefore, 2.8 million Finns lived in wooden buildings.

The latest Finnish accident statistics are for 1974. They show fatal mishaps grouped by causes as follows:

1/Ignitability class 2 = slowly igniting surface. In a test by defined procedure the surface does not ignite for at least 5 minutes.

Statistics on causes of accidental death in Finland in 1974

Transport accidents	1,113
Accidental falls	539
Accidental drowning and submersion	218
Accidental poisoning	356
Accidents caused by fire and flames	106
Accidents caused by inhalation, suffocation and ingestion	78
Other accidents	258
Total	2,668

Of the 106 deaths caused by fire, some occurred in factories, warehouses, etc. The total occurring in living quarters was 69. Regrettably the statistics do not tell us how these cases were divided between wooden and stone or brick houses.

If we compare the figure 69 with the total of fatal accidents (2,668), we obtain a percentage of 2.6. In this light it seems that fire regulations in Finland have performed their task with regard to human safety.

An interesting feature which shows up from time to time is the critical attitude of the public press toward fire accidents. This may bring pressure to bear on officials responsible for the issue and supervision of fire regulations. What is the cause of this? A full explanation is perhaps difficult, but one reason may be that the authority responsible for fire arrangements can be clearly pointed to. If, on the other hand, 20 times as many die each year in traffic accidents, or 10 times as many in falls and collapses, no particular attention is paid. For these types of accidents, however, there is no specific target for criticism.

The wood industry must, and of course will, give attention to the substance of fire regulations and their application in practice. This is necessary for the added reason that the fire characteristics of wood-based products may be wrongly understood. If fire regulations adhere to a performance concept, as in Finland, the wood industry will be given the opportunity through product development to improve not only materials but the fire-resistant properties of structures. Results are seen in increased safety for those who use wood-based products. Obviously too, the wood industry must feel responsible to society, and must seek to develop its products and wood construction in collaboration with the fire authorities in such a way that users are exposed to no unreasonable risks. The fire regulations in Finland are still so new, however, that product development has not gone far yet.

Les Risques d'Incendie — le Point de Vue des Assureurs

J.C. Roux
Comité européen d'assurances

Ce point de vue est exprimé par les dispositions prévues par les Assureurs pour la modulation du montant des primes. En assurance Incendie, la prime est le produit d'un coefficient dit taux exprimé en % par le total des capitaux garantis. Les paramètres considérés pour le calcul du taux correspondent schématiquement à quatre ensembles qui déterminent le risque d'incendie, c'est-à-dire pour les assureurs : le danger de survenance d'un incendie et, lorsqu'il s'est déclaré, le danger d'extension physique à laquelle est liée l'importance de l'indemnité versée.

Ces facteurs sont :

a) la nature précise des activités exercées, les procédés mis en oeuvre, les matières premières utilisées, etc.;

b) la qualité des bâtiments et leur conception, c'est-à-dire la nature des matériaux employés dans la construction, le compartimentage, le chauffage, l'installation électrique, la propreté des locaux et leur encombrement, les contiguités entre bâtiments, etc.;

c) les moyens spécifiques de lutte contre l'incendie;

d) la répartition des biens garantis en groupes indépendants de valeurs exposées à un même sinistre.

Le taux propre d'un bâtiment particulier est déterminé à partir d'un "taux de base" qui reflète les dangers liés à l'ensemble des facteurs a), et auquel sont appliquées des réductions et/ou des majorations qui traduisent les autres caractéristiques du risque 1/ à tarifer.

1/ En matière d'assurance, le mot "risque" a une signification très large :
- il représente d'abord le péril couvert par l'assureur(risque d'incendie, risque d'explosion, etc.);
- par extension, il indique aussi la chose sur laquelle porte l'assurance (bâtiment ou groupe de bâtiments par exemple); c'est avec ce sens qu'il figure dans les expressions :
  - "risques industriels" : établissements industriels et commerciaux ou risques qui leur sont assimilés compte tenu de l'importance des valeurs qu'ils représentent;
  - "risques simples" : bâtiments à usage d'habitation et/ou d'activités de type artisanal.

Les critères de tarification concernant la construction, qui nous intéressent ici, correspondent aux questions que se posent les assureurs lors de l'évaluation de la qualité des bâtiments vis-à-vis du risque d'incendie. Les principales de ces questions sont :

- quelle protection le bâtiment offre-t-il au contenu contre les causes extérieures d'incendie ?

- quelle aggravation peut-il apporter si le foyer initial est situé à l'intérieur du bâtiment : alimentation en éléments combustibles, effets de propagation, développement par amenée d'air ?

- le bâtiment peut-il limiter le développement de l'incendie, par exemple du fait de son compartimentage, de son incombustibilité ?

- comment est-il capable de résister, pris en lui-même, à la destruction par le feu ?

Indiquons également que les dispositions tarifaires sont définies de matière distincte suivant qu'il s'agit des "risques industriels" ou des "risques simples" pour lesquels notamment les critères d'appréciation de la construction sont moins nombreux et donnent lieu généralement à une estimation globale.

Enfin, il convient de préciser que, si bien entendu les principes généraux rappelés sommairement ci-dessus sont appliqués dans tous les pays notamment européens, leurs modalités d'application définies par les tarifs nationaux d'assurances diffèrent en général sensiblement d'un pays à l'autre.

Nonobstant cette dernière remarque, il peut être retenu qu'en ce qui concerne les "risques simples", lesdits tarifs prennent en considération d'une manière générale la nature des murs extérieurs et de la couverture des bâtiments ainsi que, parfois, celle de l'ossature porteuse. Certains tarifs modulent en outre les taux de prime correspondants selon que les bâtiments sont des maisons individuelles ou qu'ils abritent plusieurs appartements et comportent plus de deux étages. Les immeubles de grande hauteur font en général l'objet de dispositions particulières se référant à la réglementation officielle.

Dans ce cadre, l'emploi du bois entraîne l'application de taux de prime généralement très supérieurs à ceux applicables aux bâtiments en maçonnerie. En raison de l'importance prépondérante accordée par les assureurs au risque d'incendie provenant d'un feu extérieur, les bâtiments, dont notamment les murs extérieurs sont en bois, supportent dans la plupart des pays des taux plusieurs fois supérieurs aux taux appliqués aux bâtiments dont les parois extérieures sont en matériaux incombustibles

La présence du bois peut cependant ne pas être sanctionnée lorsque le bâtiment assuré :

- est séparé de toute autre construction par une distance minimale de 10 mètres (pouvant être réduite à 5 mètres si elle est libre de tout "objet")

- et possède, ainsi que les bâtiments voisins, une couverture en matériaux incombustibles.

Enfin, il convient de préciser que l'évocation du cas des "risques simples" serait très incomplète si l'existence de contrats d'assurances "multirisques" n'était pas mentionnée. En effet, ces contrats sont de plus en plus répandus et représentent certainement, dans tous les pays et dès maintenant, une part très importante du marché de l'assurance des "risques simples". Dans un même contrat, ces formules d'assurances regroupent notamment les garanties incendie, explosion, vol, dégâts des eaux, bris de glace; le risque d'incendie n'y est donc pas prépondérant. Il en résulte que pour les maisons d'habitation individuelles et les appartements assurés dans le cadre d'une police d'assurances "multirisques", les assureurs ont tendance à ne pas faire actuellement de distinction tarifaire selon la nature des matériaux de construction des bâtiments.

Cette simplification extrême, au demeurant peu justifiée au plan technique,paraît inévitable. En effet, le très grand nombre des contrats d'assurances "risques simples" impose une gestion administrative informatisée qui ne permet pas une différenciation poussée des tarifs et, surtout, la modicité de la prime moyenne desdits contrats rend impossible, en raison des frais généraux qu'elle entraînerait, la vérification de chaque risque nécessaire à la prise en considération de critères de tarification nombreux et de caractère plus technique.

Il n'en est pas de même en ce qui concerne l'assurance des "risques industriels"; dans ce domaine les assureurs ont mis au point et appliquent des dispositions tarifaires tenant compte des nombreux facteurs, déjà mentionnés, qui déterminent le risque d'incendie; une part importante de ces dispositions concerne l'influence de la construction sur les dommages incendies.

Par là même, les assureurs exercent, dans le domaine de la prévention, une influence importante en matière de conception des établissements industriels; nous verrons dans un instant quelles sont les dispositions prises concernant l'utilisation du bois et des produits à base de bois ainsi que les raisons pour lesquelles elles ont été retenues pour la tarification des "risques industriels". Toutefois, au préalable, il paraît nécessaire de préciser quelles sont les interférences de l'assurance et de la prévention car, au premier abord, il peut paraître surprenant de voir l'assureur intervenir en matière de prévention puisqu'il s'agit d'un domaine tout-à-fait distinct de sa technique professionnelle.

On pourrait concevoir, il est vrai, que l'assureur incendie soit uniquement un répartiteur d'indemnités, travaillant avec sérénité sur la base d'études techniques, notamment statistiques, et évaluant le montant des primes à percevoir suivant l'importance des risques qu'il garantit, quelle que soit cette importance. Il lui suffirait, dans cette hypothèse, de veiller à ce que ses encaissements soient à tout moment suffisants pour lui permettre de faire face à ses obligations. Selon cette conception, les problèmes de prévention sont laissés à l'initiative des pouvoirs publics et des assurés; plus précisément, les assurés sont seuls juges du degré de sécurité qu'ils entendent maintenir dans leurs biens, à l'intérieur des limites, quand il en existe, fixées par les pouvoirs publics.

Il n'est à l'évidence pas souhaitable, si tant est que cela soit possible, que les assureurs adoptent une telle attitude, et ce, notamment pour les raisons suivantes :

- il serait tout à fait regrettable que la collectivité ne bénéficie pas de la très vaste expérience qu'ils ont acquise lors des visites après sinistres et des enseignements extrêmement riches qu'ils retirent de leurs statistiques.

- même si l'assurance apporte sa répartition, les incendies se traduisent toujours au bout du compte par une perte importante pour la nation.

Au reste il convient de reconnaître que les assureurs souhaitent que leurs actions en faveur de la prévention, seront vraiment efficaces et qu'elles permettront de freiner la dégradation de leurs résultats sans qu'il soit nécessaire de recourir à de nouvelles augmentations des primes. Par ailleurs, l'effort de prévention est devenu une nécessité pour certaines catégories de risques, car les majorations dont elles seraient normalement passibles constitueraient des charges supplémentaires difficilement supportables par les assurés.

Notre propos n'est pas ici d'expliquer de quelle façon les assureurs encouragent la prévention, mais, la politique tarifaire étant le moyen le plus efficace dont ils disposent pour la promouvoir, nous connaîtrons quelques uns des moyens utilisés en examinant les dispositions tarifaires concernant la construction applicable aux "risques industriels".

Tous les tarifs nationaux d'assurance incendie considèrent la nature des matériaux constitutifs des murs extérieurs et de la couverture ainsi que la stabilité au feu de l'ossature générale des bâtiments. Les aménagements intérieurs tels que notamment les sous-toitures, les faux plafonds, les sous-plafonds et les cloisonnements verticaux, sont également pris en considération par la plupart desdits tarifs. Enfin, ceux-ci tiennent compte aussi parfois d'aménagements particuliers, il en est ainsi par exemple, de la présence d'exutoires de fumée et/ou de dispositifs d'évacuation de l'eau utilisée pour l'extinction par les pompiers.

Les modalités d'application des dispositions tarifaires concernant ces diverses parties de bâtiment diffèrent, ainsi que nous l'avons déjà dit, sensiblement d'un pays à l'autre. Toutefois, les méthodes de prise en considération des bâtiments peuvent être sériées en deux catégories principales.

La première de ces méthodes consiste à définir plusieurs types de bâtiments de comportement différent au feu. Le tarif indique alors pour chacun de ces types un coefficient d'appréciation globale de son influence sur le risque d'incendie.

La simplicité de cette méthode n'est en fait qu'apparente puisque l'inspecteur d'assurance se trouve le plus fréquemment en présence d'un bâtiment qui n'est entièrement conforme à aucun des types de bâtiments prévus. Il en résulte qu'il doit interpréter les dispositions du tarif alors que sa compétence en matière de comportement au feu peut être insuffisante à cet égard.

Au contraire la seconde méthode, utilisée sensiblement par un même nombre de pays que la première, consiste à prendre en considération séparément chacune des parties de bâtiment retenues comme ayant une influence sur les dommages d'incendie. Cette méthode semble mieux adaptée à l'extrême variété des bâtiments à assurer puisqu'elle permet de déterminer le tarif correspondant à des combinaisons multiples. Il lui est toutefois reproché d'être d'une application assez complexe et de ne pas présenter un caractère suffisamment didactique pour les constructeurs qui souhaiteraient connaître les dispositions recommandées par les assureurs. C'est pourquoi, plusieurs pays ont complété cette méthode en définissant également un nombre restreint de types de bâtiments donnant lieu à l'application de rabais de prime importants. Les définitions correspondantes comprennent des spécifications relatives tant à la conception de l'immeuble qu'à sa construction proprement dite lui assurant une "résistance au feu" généralement d'au moins 1 h, parfois plus.

Il convient de préciser que ces exigences importantes des assureurs concernant la stabilité au feu de bâtiments industriels donnant lieu à l'application d'un rabais de prime tiennent à deux raisons essentielles : la gravité des incendies industriels et la recherche de la protection des biens matériels.

La violence et l'extension d'un incendie sont, notamment, fonction de la qualité et de la nature des marchandises et du matériel qui se trouvent dans les locaux atteints. En outre, le potentiel calorifique généralement important du contenu des locaux industriels se traduit notamment par une augmentation de la durée de l'incendie.

Il faut également considérer les températures qui peuvent être atteintes au cours d'un incendie industriel. La courbe température/temps reconnue internationalement a été établie à partir de relevés de températures effectués lors d'incendies réels de locaux généralement à usage d'habitation. D'après cette courbe par exemple, une température de 500°C est atteinte en quelques minutes et après une 1/2 heure la température dépasse 800°C pour tendre ensuite vers 1200°C. Ce sont là des températures importantes; or, dans un incendie industriel, de plus en plus souvent, les températures sont beaucoup plus élevées et parviennent à ces valeurs beaucoup plus rapidement.

Ces caractéristiques des incendies industriels sont dues essentiellement à la présence dans les bâtiments de matières dont la combustion s'accompagne de débits calorifiques extrêmement élevés. Il s'agit notamment de solvants dont l'utilisation est de plus en plus fréquente dans les ateliers et des matières plastiques qui sont utilisées aujourd'hui non seulement dans les emballages et dans de nombreux produits manufacturés mais aussi dans les équipements et bâtiments à usage industriel. La pyrolyse de ces matières organiques dégage également :

- soit d'énormes quantités de fumées qui constituent une gêne pour les secours et souvent même rendent impossible toute intervention manuelle efficace sur les foyers;

- soit des gaz combustibles qui, s'ils ne trouvent pas l'oxygène nécessaire à leur combustion là où a débuté l'incendie, se répandent à grande distance du lieu de l'origine du sinistre, provoquant des extensions inattendues du feu et parfois des explosions. Il faut voir là une explication de la violence des grands sinistres, violence considérée souvent comme mystérieuse.

Pour ce qui est de la recherche de la protection des biens matériels une remarque s'impose. L'évacuation des biens n'étant généralement pas possible, la stabilité au feu du bâtiment doit être particulièrement recherchée puisqu'en dépendent notamment les conditions d'intervention des secours.

En outre, par exemple, au cours d'un incendie, tant que la toiture reste en place, il y a des possibilités de tenir le feu en échec alors que si la couverture s'effondre, le sinistre est généralement total.

Revenons aux dispositions tarifaires appliquées par les assureurs pour préciser qu'une comparaison des coefficients de variation de la prime recommandés par les différents tarifs nationaux pour un même type de bâtiment n'a pas de réelle signification. En effet, les valeurs retenues sont nécessairement assez différentes car lesdits coefficients s'appliquent à un taux de "base" qui reflète statistiquement les dangers propres à l'activité exercée et qui caractérise le "comportement d'assurance incendie", pendant une période donnée, d'un ensemble de risques renfermant la même activité. Ce comportement statistique dépend donc, entre autres choses, du type de bâtiment le plus répandu dans ladite activité industrielle. Or, ce type de bâtiment "standard" n'est la plupart du temps pas identique d'un pays à l'autre notamment dans le cas des immeubles anciens 1/ qui constituent une part généralement importante du parc des bâtiments existants. Le fait qu'en pratique les taux "de base" indiqués dans les tarifs ont été, au fil des années, progressivement corrigés afin de permettre l'application de coefficients "pour construction" identiques, quelle que soit l'activité considérée, n'enlève rien à la généralité de cette remarque.

Ceci étant, concernant la difficulté de comparer valablement les dispositions tarifaires nationales relatives à la construction des "risques industriels", il paraît préférable d'analyser ici les dispositions d'un de ces tarifs. Appartenant moi-même à l'Assemblée Plénière des Sociétés d'Assurances contre l'Incendie (A.P.S.A.I.), qui est l'association regroupant les Sociétés pratiquant l'Assurance Incendie en France, c'est bien entendu le "Tarif des Risques Industriels" établi par cet organisme que je connais le mieux, ce sont donc ses dispositions que je commenterai.

Ce tarif classe les matériaux utilisables pour la construction des murs extérieurs, des couvertures et des aménagements intérieurs (contre-murs, sous-toitures, faux-plafonds et sous-plafonds) en trois catégories : les matériaux "durs", les matériaux "semi-légers" et les matériaux "légers". Ce classement est établi en fonction, notamment, des quantités de résistance au feu des matériaux mais d'autres critères sont également pris en compte : pouvoir calorifique, toxicité et corrosivité des gaz de combustion, dégagements de fumée, etc.; la meilleure catégorie est celle des matériaux "durs".

La méthode de "classement" des bâtiments utilisés est celle qui consiste à considérer séparément chacune des parties de bâtiment. Ainsi, les murs extérieurs et les couvertures sont "classés" d'après le pourcentage des surfaces occupées par les matériaux des différentes catégories (durs, semi-légers, légers) par rapport à la surface respective totale desdits murs et des couvertures (cf. le tableau des majorations, pour construction mixte ou légère, dépendant de la nature des murs extérieurs et la couverture).

1/ On constate, par contre, une homogénéisation progressive des constructions industrielles récentes, et ce, quelque soit l'activité exercée.

Par ailleurs, selon leur nature, les contre-murs d'une part, les sous-toitures et les faux-plafonds d'autre part peuvent modifier respectivement le "classement" des murs extérieurs et de la couverture : les sous-plafonds éventuels sont pris en compte avec les planchers classés en fonction de leur résistance au feu en deux catégories.

Dans le cas général des constructions ne présentant pas une "résistance au feu" supérieure à environ une heure, qui font l'objet de définitions particulières, il n'est pas fait de distinction entre les ossatures métalliques, en béton armé ou en bois. Toutefois, il faut pour cela que les ossatures en bois répondent aux prescriptions du "Cahier de Spécifications C.S.1" établi par l'Assemblée Plénière et l'organisme professionnel regroupant les entrepreneurs de charpentes.

Ainsi, alors que tous les éléments de construction en bois sont considérés comme constitués de matériaux "légers", une ossature en bois lamellé-collé ou en bois massif conforme audit "Cahier C.S. 1" est assimilée à une ossature en matériaux "durs" et ne donne lieu à l'application d'aucune majoration de prime. La prescription essentielle du "Cahier C.S. 1" concerne le respect de sections minimales pour les éléments d'ossature principale et les éléments complémentaires tels que les lisses et les poteaux de bardage; d'autres prescriptions concernent notamment la protection vis-à-vis du feu des organes d'assemblage de ces divers éléments.

Parmi les couvertures il existe également un cas particulier pour lequel le bois n'est plus considéré comme "léger" mais comme matériau "semi-léger". Il s'agit des "couvertures à supports d'étanchéité en bois, panneaux dérivés et assimilés" conformes au "Cahier de Spécifications C.B. 2" établi avec des fabricants et des entrepreneurs groupés sous l'égide de la Fédération Nationale du Bâtiment.

Ce type de couverture est constitué des éléments suivants (de l'intérieur vers l'extérieur) : un support ligno-cellulosique rigide, un isolant thermique (éventuellement), un revêtement d'étanchéité. Outre des prescriptions générales de composition et de pose des supports autorisés 1/, les éléments constitutifs de la couverture doivent présenter les caractéristiques minimales de comportement au feu définies dans le tableau ci-après.

Par ailleurs, il convient de signaler que l'un et l'autre de ces Cahiers de Spécifications prévoient la délivrance par l'entrepreneur, pour chaque bâtiment, d'un certificat de conformité auxdites prescriptions. Destinée en premier lieu à faciliter l'identification des matériaux par les inspecteurs d'assurance, qui se font présenter le certificat par l'assuré, lorsqu'ils vérifient les "risques industriels", cette procédure est suivie, pour certaines familles de matériaux, depuis plusieurs années à la satisfaction de tous.

A ce stade, il me paraît nécessaire de rappeler que, par convention, le risque ordinaire est celui dont aucune des caractéristiques, autre éventuellement que l'activité industrielle qui y est mise en oeuvre, ne constitue

---

1/ contreplaqués multiplis, panneaux de particules de bois, de fibres de bois agglomérées par un liant hydraulique, d'anas de lin, panneaux silico-cellulosiques (paille) et platelages composés de planches rainées bouvetées (type parquet).

<table>
<tr><td colspan="2" rowspan="2">TYPE DE COUVERTURE</td><td colspan="2">SUPPORT</td><td rowspan="2">REVETEMENT D'ETANCHEITE</td></tr>
<tr><td>Résistance au feu
Degré coupe-feu minimum</td><td>Réaction au feu</td></tr>
<tr><td colspan="2">Sans isolation thermique</td><td>1/2 H</td><td rowspan="3">M3 minimum</td><td rowspan="3">Le revêtement d'étanchéité associé au support doit remplir au minimum l'une des deux conditions suivantes :
- être classé M0, M1, M2 ou M3
- avoir obtenu (associé à son support et à l'isolant thermique éventuel) le classement T 30/1</td></tr>
<tr><td rowspan="2">Avec isolation thermique intégrée ou rapportée</td><td>P.C.S. de l'isolant inférieur ou égal à 2000 cal/g</td><td>1/2 H</td></tr>
<tr><td>P.C.S. de l'isolant supérieur à 2000 cal/g mais inférieur ou égal à 4000 cal/g</td><td>1 H</td></tr>
</table>

un danger grave d'incendie ni une mesure de prévention très efficace; la prime correspondante ne dépend donc que de l'activité. Le "Tarif des Risques Industriels" précise quelles sont, pour chacun des facteurs pris en considération dans l'évaluation du danger d'incendie, les normes du risque ordinaire.

Pour ce qui est de la construction, le bâtiment "ordinaire" est un immeuble dont les éléments essentiels (structure portante, murs, couverture, plancher et aménagements intérieurs) sont en matériaux "durs" avec tolérance de quelques pour cent de matériaux de caractéristiques inférieures - c'est-à-dire, une construction neutre ne contribuant pas à la propagation de l'incendie mais ne présentant pas toutefois la meilleure stabilité possible en cas d'incendie d'envergure.

Nous pouvons donc conclure que le "Tarif des Risques Industriels" de l'A.P.S.A.I. pénalise l'utilisation en proportion importante - supérieure à une tolérance générale d'emploi de 10 % (cf. tableau des majorations pour construction mixte ou légère) - du bois dans la construction des bâtiments mais que, sous réserve du respect de certaines spécifications au demeurant peu contraignantes, cette pénalisation peut être fortement réduite voire supprimée pour ce qui est des structures.

MAJORATIONS POUR CONSTRUCTION MIXTE OU LEGERE.

MURS EXTERIEURS (Dur / Semi-Léger / Léger)	COUVERTURE 1re CLASSE — toute nature avec : Plus de 90 % de mat. durs	2me CLASSE — toute nature avec : Moins de 10 % de mat. légers	3me CLASSE — toute nature avec : Plus de 10 % de mat. légers
	%	%	%
1er RISQUE Plus de 90 % de matériaux durs	0	20	50
2me RISQUE De 10 à 50 % de matériaux légers et semi-légers, mais moins de 10 % de matériaux légers	20	40	70
3me RISQUE Plus de 10 % de matériaux légers ou plus de 50 % de matériaux légers et semi-légers	50	70	100

Les assureurs sont favorables à une collaboration avec les professions intéressées du bâtiment et souhaitent que celle-ci se poursuivre afin que soient mises au point des dispositions améliorant le comportement au feu des bâtiments afin qu'ils puissent promouvoir ces dispositions préventives, éventuellement au moyen d'une incitation tarifaire.

Insurance Rates for Wooden Houses

J.T. Wassink
Royal Tropical Institute, Netherlands

SUMMARY

A STRANGE SITUATION IN BELGIUM, ENGLAND, FRANCE AND THE NETHERLANDS

In table 1 (bilingual table in French text) fire insurance rates are given for houses in various construction categories (stone, wood, hard roof, soft roof) for a great number of countries.

The type of house we have chosen is the normal dwelling house, situated in normal surroundings without special risks. For some countries, which have forested regions as well as regions without forests, several data have been gathered.

It appears that in countries (or regions) where wood is traditionally used for building purposes the rates are lower, and indeed often very much lower, than in the Netherlands, Belgium, France and England. The insurance companies do not seem to go bankrupt on this system.

A piquant detail is that Canada, with its own low rates, re-insures in England where the rates are high for wooden houses.

In Sweden, the rates for different types of construction systems have been levelled for the last 5 years (which means less work for the insurance people) and it seems that this works out reasonably well.

As samples taken at random, Arnhem (Netherlands) masonary-built and Paramaribo (Surinam), of the same size but built of wood, give the following results:

Masonary-built town (99% brick), mean number of serious fires during 5 years 7.6, alarms given 344. Wooden town (more than 60% wood), mean number of serious fires in 21 years 5.5, alarms given 44.4.

These data have been brought to the attention of the Minister of Housing in the Netherlands, with the idea of taking away one of the obstructions to building in wood.

Primes d'Assurance Incendie pour les Maisons d'Habitation en Bois

J.T. Wassink
Institut Royal des Tropiques, Pays-Bas

UNE SITUATION ETRANGE AUX PAYS-BAS, EN BELGIQUE, EN FRANCE ET EN ANGLETERRE

Des primes d'assurance incendie élevées constituent une entrave à l'utilisation du bois dans la construction immobilière et les instances de contrôle de la construction se réfèrent à ces primes élevées pour démontrer le danger de construire des maisons en bois.

L'énoncé ci-dessus est une histoire bien connue qui a amené l'occasion de beaucoup de discussions et de recherches dans le domaine de la technologie du feu.

"La construction en bois est-elle ou non dangereuse par le feu?"

Ci-dessous, ce point est analysé sous un autre aspect. L'Europe (et le monde) sait discerner où la construction de maisons en bois est normale et d'autres pays, comme les Pays-Bas où la construction de maisons en bois constitue un fait unique.

Comment cela se passe-t-il pour les primes d'assurance-incendie?

L'auteur de cet article a demandé à un grand nombre de pays le montant des primes d'assurance contre l'incendie pour les maisons d'habitation dont la situation ne provoque pas un risque d'incendie anormal, ainsi que les primes pour les méthodes de construction ci-après :

1) Maison de pierres avec un toit en dur (tuiles, ardoises, etc.)

2) Maisons en bois avec un toit tendre (paille, bardeaux, papier asphalt).

Il va de soi que d'autres facteurs peuvent influencer la prime de base: par exemple le système de chauffage.

Le point 3 a soulevé quelques difficultés : les assurances à l'étranger ne font souvent aucune différence, entre les toits durs ou tendres, ou bien le toit tendre est totalement inconnu.

L'enquête en question est donc limitée mais on a demandé le montant des primes aux pays qui ont peu de forêts et à ceux qui en ont beaucoup.

Les résultats de ces enquêtes figurent au tableau 1.

COMMENTAIRES DU TABLEAU 1

Maison en pierres - toit en dur

Les primes diffèrent d'un pays à l'autre, mais l'amplitude, exception faite du Canada et de l'Etat le plus cher des Etats-Unis, est relativement

Tableau 1 - Table 1

Primes d'assurance incendie dans certains pays
Fire insurance premiums in selected countries
(Taux pour mille/Rate per mil)

Matériau de construction Toit Building material Roof	brique dur stone hard	bois tuile wood hard	bois tendre wood soft
Autriche/Austria b/	0,112	1,2	1,6
Allemagne, Rép. féd.d'/ Germany, Fed. Rep. of, Hamburg b/	0,21	1,25	1,59
Allemagne, Rép. féd. d'/ Germany, Fed. Rep. of, Karlsruhe b/	0,22	0,43	0,43
Allemagne, Rép. féd. d'/ Germany, Fed. Rep. of, Munich b/	0,22	0,43	0,55
Autriche II/Austria II b/	0,23	0,96	1,6
Finlande/Finland	0,3	0,4	0,5
Autriche III/Austria III b/	0,32	0,96	1,6
Etats-Unis (47 états) Min./ USA (47 states) Min. a/	0,47	0,65	+ max. $ 3.-
Pays-Bas/Netherlands	0,5	3,0-5,0	5,0-10,0
Autriche IV/Austria IV b/	0,56	1,7	2,8
Suisse (Baselland)/Switzerland (Baselland)	0,6	0,8	?
Canada S. Ontario	0,6	0,45-1,05	?
Belgique/Belgium	0,55-0,65	3,3	6
Suède/Suède b/	0,5-0,7	0,5-0,7	?
Norvège/Norway	0,7	0,8	0,8
Angleterre/England	0,75	2,5-3,7	?
Suisse/Switzerland Graubünden	0,8	1,2	2
Canada N. Ontario	0,8	0,5-1,25	?
Etats-Unis (47 états) moyen/ USA (47 states) mean a/	1,13	1,5	+ max. $ 3.-
Etats-Unis (47 états) max./ USA (47 states) max. a/	2,47	2,65	+ max. $ 3.-
France	?	3,0	6

a/ Fire protection class 3. There are ten such classes in the USA. Class I has a perfect fire protection organization; Class 10 has no protection whatsoever. Class 3 is chosen for reasons of comparison.

Classe de protection contre l'incendie 3. Il existe 10 de ces classes aux Etats-Unis. La classe I a une organisation parfaite de protection contre l'incendie. La classe 10 n'a aucune protection. La classe 3 a été choisie pour des raisons de comparaison.

b/ Extended coverage for hail, storm and water damage, as well as fire

Couverture étendue pour les tempêtes de grèle, les dégats d'eau ainsi que les incendies.

mince. Les Pays-Bas prennent au milieu de la série une place assez raisonnable tandis que les différences avec la Belgique et le Royaume-Uni ne sont pas sensationnelles.

Maison en bois avec un toit en dur

L'amplitude des primes est ici un peu plus nuancée. Il est surprenant que quatre pays européens (Belgique, France, Pays-Bas et Royaume-Uni) et le Texas sont en tête. Il s'agit de pays relativement peu boisés. Près de la moitié des pays producteurs de bois ont des primes qui se trouvent à un niveau notablement bas, à savoir 1/6 environ inférieur aux autres. Dans ces "pays bon marché", la construction en bois est une affaire acceptable. Les exercises financiers des assurances de ces pays sont meilleurs que par exemple aux Pays-Bas et en Angleterre.

Il est étonnant que le Canada, avec une amplitude de prime de 0,45 à 1,25 %, se réassure en Angleterre où les primes locales sont beaucoup plus élevées, entre 2,5 et 3,7 %.

En Suède, depuis quelques années, on ne fait plus aucune différence entre le bois et la pierre en tant que matériau de construction; une des raisons de cette situation était l'économie de coûts de personnel recherchée par les compagnies d'assurance. L'expérience des années passées (environ cinq) n'a pas encore donné de raisons pour amener un changement.

Lorsqu'on examine le tableau, en observant que d'autres pays ont beaucoup plus d'expérience avec l'assurance de maisons en bois, une prime de 1,5% aux Pays-Bas pourrait être un maximum pour ce type d'assurance au lieu de 3 à 5 % qui sont actuellement en vigueur.

Maison en bois avec un toit tendre[1/]

Ce qui s'applique ci-dessus pour les maisons en bois est valable aussi pour ce groupe; pour les Pays-Bas, une prime de 2 % maximum devrait être appliquée au lieu de 5 à 10 % appliqués actuellement.

L'auteur sait parfaitement que les grandes compagnies peuvent pratiquer des primes relativement meilleur marché que celles citées dans le texte.

Cela nous amène au point de remarquer qu'en matière d'assurance incendie, on est vite perdu mais cette méconnaissance se situe surtout dans le domaine de l'assurance industrielle. Dans ce domaine, les compagnies d'assurance fixent en fonction des primes les parties de risques qu'elles acceptent de couvrir.

Les Pays-Bas, au cours de ces dernières années, ont connu quelques incendies sérieux. Il ne paraît pas juste à l'auteur que le règlement de ces affaires se répercute sur le petit propriétaire (souvent privé) de maison en bois.

Villes en pierres comparées aux villes en bois

A la lumière des curiosités observées ci-dessus, l'auteur a comparé deux grandes villes à peu près similaires. Il a coisi deux villes qu'il

1/ Aux Etats-Unis, indépendamment du montant à assurer, une prime complémentaire de US $ 3 est demandée pour les maisons qui sont couvertes de bardeaux; au-delà, le toit n'influence pas la prime.

connaît personnellement, l'une avec beaucoup de constructions en bois, Paramaribo, l'autre, sans constructions en bois, Arnhem.

Pour ces deux villes, il n'a pas pu trouver un critère pour déterminer le nombre total de constructions actuelles, il reste la division entre le matériel utilisé pour la construction. Pour ce faire, il a utilisé les impôts payés.

Les données statistiques se rapportant au feu et au danger d'incendie figurent au tableau 2.

Tableau 2

Comparaison des villes de Paramaribo et Arnhem

	Arnhem	Paramaribo
Total habitants	132531 (1.2.70)	± 140000
Total habitations	?	?
dont en bois	négligeable	+ 60 %
Industrie	beaucoup	peu
Incendies:		
chiffres moyens pour la période Par année	1965-1969 (= 5 ans)	1949-1969 (= 21 ans)
Total alarmes incendie	344	44,3
Total incendies graves	7,6 (min.7 max.9)	5,5
Construction en bois (inc. grâve)	0,6	
dont : maisons d'habitations	0,0	
maisons de pierres, toit en chaume	0,4	
Constructions en bois (tous incendies)		13,5
Constructions en pierres " "		1,1
Constructions en fer " "		1,0
Constructions en acier " "		5,2
Constructions en aluminium " "		0,1
		20,9

Les données se rapportant à la ville d'Arnhem ont été fournies par le commandant des pompiers, celles de Paramaribo par le Commissaire du District de Paramaribo. Les constructions en bois qui ont brûlé à Arnhem étaient entre autres une grange et une école d'assistance.

La statistique pour Paramaribo montre que le pourcentage d'habitations en bois qui ont brûlé correspond au pourcentage de constructions en bois dans la ville. Les chiffres ne permettent pas d'indiquer si une ville en bois est plus dangereuse sur le plan incendie qu'une ville en pierres.

L'auteur est plus convaincu que le danger d'incendie d'une habitation est déterminé par le contenu de l'habitation et que le dommage qui résulte de l'incendie dépend de l'intensité du feu, indépendamment du matériel utilisé pour la construction. L'auteur a entrepris des démarches auprès des assurances contre l'incendie afin qu'ils soient au courant de l'information rassemblée.

Dans la pratique, il apparaît qu'aucune instance ne se sent préoccupée par cette matière. La seule instance qui ait accepté une entrevue a été le Secrétariat des Assurances contre les incendies dans l'Industrie, à Bilthoven, mais le sujet n'a pas été abordé en détail.

Il est apparu à l'auteur qu'il n'existe aucun lien d'analogie entre les primes pratiquées par les compagnies pour ce genre d'assurance.

De droit, chaque compagnie peut faire ce qu'elle veut quant aux primes, et cela sans risquer d'être traitée d'"outsider".

Activities on International Harmonization of Building Codes and Regulations

Environment and Human Settlements Division of the Economic Commission for Europe 1/

At the conclusion of the fourth Seminar on the Building Industry (London, October 1973), organized by the Committee on Housing, Building and Planning of the United Nations Economic Commission for Europe, a Policy Statement was made on the need to harmonize building laws, regulations and standards. 2/

Today it is recognized that the differences between them constitute an obstacle to international trade as important as customs barriers and one which is much more difficult to eliminate.

On the basis of the Policy Statement, the Working Party on the Building Industry adopted at its seventh session (March 1974) a long-term programme of work on the harmonization of the technical content of building regulations in the ECE region.

This programme is made possible by contributions in cash or in kind made by most of the countries represented in ECE.

The objective of the fourth part of the long-term programme on the harmonization of the technical content of building regulations is to arrive at specific proposals for harmonization. Accordingly, rapporteurs were instructed to undertake a critical review of work in progress in intergovernmental sub-regional and international non-governmental organizations.

For project 4.3, these experts are Mr. Steckhan (German Democratic Republic) and Mr. Wahlström (Sweden).

In collaboration with the Secretariat, these experts submitted report HBP/WP.2/R.74 of 27 October 1976 (attached), to the ninth session of the Working Party on the Building Industry.

At its ninth session in December 1976, the Working Party on the Building Industry took note of it, and of the report of the WG (Working Group) 14 "Fire" of C.I.B. (the International Council for Building Research, Studies and Documentation), which is annexed.

The Governments and the non-governmental technical organizations concerned have been requested to make their views known on this report and on the changes required to bring it up to date.

1/ Valuable assistance was given in preparing this report by Mr. G. Blachère, Secretariat Consultant

2/ HBP/SEM.3/CRP/13.

The Working Party decided to hold, during the year 1977, an inter-secretariat meeting at the ECE Secretariat of two sub-regional governmental organizations and non-governmental organizations concerned, to have an exchange of information regarding intentions, achieve optimum utilization of the technical documents produced by the international technical organizations and expedite harmonization of regulations.

It is generally recommended that every effort should be made to bring about harmonization of regulations by utilizing regulations based on user requirements (e.g. functional requirements) rather than on technical specifications which call for detailed solutions.

Unfortunately, owing to insufficient knowledge in the field of fire safety, this is not yet possible.

It will be noted from the report that the regulations now under consideration contain no trace of discrimination regarding materials.

This ought to lead to the elimination of prejudices and practices concerning the use of certain inflammable materials such as wood and wood products.

HARMONIZATION OF THE TECHNICAL CONTENT OF BUILDING REGULATIONS

PROJECT 4.3 - FIRE PROTECTION REQUIREMENTS FOR RESIDENTIAL BUILDINGS

Report[1/] prepared by Mr. B. WAHLSTRÖM (Sweden), rapporteur, in consultation with Mr. P. STECKHAN (German Democratic Republic), rapporteur, and the secretariat 2/

INTRODUCTION

1. It will be recalled that at its eighth session, June 1975, the Working Party on the Building Industry decided *inter alia* to:

> (b) "Request the secretariat to keep informed of the activities of, and maintain liaison with, sub-regional intergovernmental organizations and non-governmental organizations in this field and ensure that relevant documentation and information is exchanged between them and the rapporteurs;

1/ Presented to the Working Party on the Building Industry of the ECE Committee on Housing, Building and Planning, ninth session, 6-10 December 1976 (document HBP/WP.2/R.74).

2/ Valuable assistance was given in this work by Mr. G. Blachère, consultant.

(c) Request the rapporteurs, in co-operation with the secretariat, to prepare substantive proposals for further action for examination at the next session of the Working Party, including proposals for the organization of a possible ad hoc meeting of experts on the subject of fire protection regulations with participation of sub-regional intergovernmental and international non-governmental organizations concerned;

(d) Endorse the decision by the Liaison Committee of CEB, CIB, ECCS, FIP, IABSE, IASS and RILEM, that the CIB Working Commission W14 should undertake a critical examination of the technical basis of proposed fire codes and establish basic rules for safety. The Liaison Committee should be invited to submit a report on the progress of this work to the secretariat by 31 July 1976 for examination by the Working Party at its ninth session;" (HBP/WP.2/6, para. 18)

2. The secretariat and its consultant have been in touch with a number of international organizations working in the fire-protection field. At the inter-secretariat meeting in March 1976, an early draft of the substantive proposals called for by the Working Party was discussed. At this meeting it was particularly suggested that attention be given to the draft standard for sprinkler systems being prepared by the European Fire Insurance Association. It was considered useful that a restricted inter-secretariat meeting be held to promote closer co-operation between CIB, ISO and CMEA, NKB and EEC in this field. Such a meeting could do much to introduce a pace more in line with the urgency of the work being accorded by sub-regional, intergovernmental organizations. The meeting felt that the most important studies in connexion with this project seemed to be those leading to international agreements on methods of test and calculation since these vary considerably from country to country.

3. The rapporteurs have met, in consultation with the secretariat on 21-24 January 1976 in Weimar and 6-7 September and 5-8 October 1976 in Stockholm in order to prepare their report to the Working Party on the Building Industry in accordance with paragraph 18 (c) of the Report of the eighth session of the Working Party.

4. In response to 1(d) above, the secretariat has received from CIB a report of the current activities of its Working Commission W14 (see annex I) and a draft of its "Principles and Guidelines for international codes for structural fire protection" currently under preparation (annex II).

5. An attempt has been made in this report to divide into various chapters the regulations which are (or might be) laid down by Governments to ensure safety in the event of fire, to highlight in each chapter the present state of the work being carried out by sub-regional intergovernmental groupings and non-governmental organizations, and to deduce from it general and specific ideas on the way in which harmonization might be achieved.

6. In the light of the programmes prepared by NKB and CMEA, and of the information available on the topic, the following chapters are proposed :

(a) Terminology

(b) Basic fire protection requirements

(c) Evacuation requirements

(d) Fire compartment requirements

(e) Requirements concerning the fire resistance of load-bearing structures

(f) Limitation of the diffusion of smoke and gas

(g) Methods of testing fire resistance

(h) Methods of testing

inflammability
the spread of fire, and
combustibility

(i) Methods of testing toxicity

(j) Methods of testing smoke capacity

(k) Automatic alarm and fire extinction equipment.

TERMINOLOGY

7. It would be useful to agree on terminology as different meanings are attributed to the various terms used in fire safety matters. ISO has prepared a preliminary draft (ISO/DIS 3261) which is supplemented as necessary. CMEA has also prepared a draft terminology. It would be helpful to compare these two drafts, and all member governments of ECE should be encouraged to use ISO terms as far as possible.

BASIC FIRE PROTECTION REQUIREMENTS

8. At present, with very few exceptions, the safety regulations in force, as well as those in preparation, make it compulsory to utilize certain means (i.e. materials, components and layouts which are described in terms of their physical characteristics, their classification according to tests, their shapes or dimensions. Such regulations fail to indicate, even for information purposes, the principles on which their safety provisions are based. This makes them difficult to apply, particularly where innovations are concerned.

9. On the other hand, the fact that such principles have not yet been incorporated in regulations suggests that it will be possible to enunciate them for the first time at the international level and thus establish a uniform basis for national or regional regulations. These principles imply, first of all, that safety depends on :

Prevention, that is, reducing the probability of a fire occurring and, if a fire does occur, reducing its consequences;

Evacuation of occupants outside the premises or to places of safety; and

Extinction of the fire by automatic means or by the fire brigade.

10. In other words, safety depends on the efforts made to reduce the likelihood of fires, reduce their severity and enable occupants to escape.

11. Evacuation will be possible if a proper relationship is maintained between certain periods of time :

The alarm time, i.e. the time that elapses between the initial outbreak of the fire and the moment the occupants become aware of it ;

The evacuation time ;

The length of time that evacuation routes remain safe ;

The length of time that places of refuge remain safe ;

The survival time, after which it is dangerous or fatal to remain in the burning premises ;

The length of time that the burning structures remain unaffected ; and

The time taken to extinguish the fire.

12. A definition of these times (unquantified) has, in fact, been proposed to ISO/TC 59/SC 3, which is responsible for the standardization of requirements. It is hoped that a draft can be put to the vote at the beginning of 1977.

13. This problem has not yet been tackled by the other organizations apart from CIB, which took it up a few years ago and is now preparing a report on the principles of structural fire protection (annex II).

EVACUATION REQUIREMENTS

14. These requirements concern the routes by which occupants can leave the premises in the event of fire and reach a place of safety. They may relate to the number, width and arrangement of doors ; the width and surfacing of corridors ; and the width, steepness, surfacing and arrangement of staircases.

15. CMEA is at present studying this subject and its principles on evacuation routes should be submitted in August.

16. Within NKB this problem has been taken up and an investigator is now working on this project and will present a proposal in 1977. For one-family houses, prescriptions within this field are already laid down.

17. It would seem that, except perhaps in the case of prefabricated houses, standardization of evacuation provisions is not essential for international trade in building materials and components.

FIRE COMPARTMENT REQUIREMENTS

18. These regulations specify the volume (or area) of the compartments comprising a building in the light of various parameters (combustible load, density of occupation and fire hazard), the compartment being defined as an enclosed space the walls of which behave in a certain manner when affected by fire.

19. The NKB and CMEA programmes include work on compartments :

CMEA is to approve a draft in August 1976 ;

NKB has completed a draft for one-family houses and, on the basis of that draft, is preparing another for apartment houses ;

In EEC, a working group is preparing a working paper which will be available at the end of 1976 ;

Neither ISO nor CIB is active in this field.

The comments made regarding evacuation in paragraph 10 above are also applicable to compartments.

REQUIREMENTS CONCERNING THE FIRE RESISTANCE OF BEARING STRUCTURES

20. The fire resistance of bearing structures has so far been determined by the resistance test, which is referred to in paragraph 32 below. In recent years, however, attempts have been made in several quarters to devise methods of calculating the fire resistance of bearing structures. Except in France, these methods have not yet been incorporated into standards or regulations. The time is therefore ripe for an attempt at unification.

21. It will be noted that CIB is responsible for the co-ordination of the work of various international organizations in this field through the Code Advisory Panel (CAP) of CIB/W 14, which intends to prepare documents on general principles in the light of the CEB/FIP document (see annexes I and II). EEC is prepared to accept the CIB proposals.

22. In June 1975, CEB and FIP published "Design recommendations for reinforced and prestressed structural members with regard to fire resistance". These recommendations are to be revised in 1976. ISO has yet to include the subject in its programme. The regional organizations NKB and CMEA have no projects in this sphere. EEC is to produce a working paper towards the end of 1976, based on the CEB/FIP document.

23. Lastly, the Working Party on the Building Industry could, as regards this subject, ask CEB and FIP to prepare a working document as soon as possible (if this is not done in 1976), which ISO could publish as a draft standard, and suggest that Governments and regional organizations should take into account, in their regulations, calculations made according to the CEB/FIP method.

24. As regards steel, ECCS is to submit recommendations at the end of 1976, as will EEC on the same basis. No other intergovernmental or technical organizations have embarked upon work in this field.

25. The same approach could be adopted as in the case of concrete, with ECCS playing the part of CEB/FIP. It would be advisable to associate CIB with this work so as to ensure that all countries are properly consulted. The Working Party on the Building Industry could inform ECCS and CIB of its hope that the CIB/ECCS calculating rules will be submitted during 1977. It could recommend that Governments and regional organizations should take these rules into account in their regulations.

26. CIB Commission W 18 is preparing rules for the calculation of wooden structures. These rules will be standardized by ISO/TC 129. It will be noted that the ECE Timber Committee intends to arrange a symposium on this question in early 1977. The regional organizations are not active in this field.

27. It appears that there is no urgency as regards other materials, except for plastics structures.

28. In addition to testing and calculation methods, there are the Regulations on fire resistance. ISO and CIB are not engaged in any work on this subject. EEC is active in this field and is preparing general fire resistance regulations ; CMEA is not giving priority to the subject and NKB is dealing with it from the standpoint of one-family houses. It is a subject not without importance for the trade in components. The Working Party on the Building Industry might therefore recommend studies which, even if they fail to specify such fire resistances, might at least give a range of resistances that Governments could use for each component. ISO TC 59 or 92 could be asked to carry out this work.

LIMITATION OF THE DIFFUSION OF SMOKE AND GASES

29. The presence in buildings of many vertical shafts and ducts (staircases, lifts, ventilation, etc.) and of horizontal communciations (corridors, ceiling cavities, ventilation, etc.), together with the greater harmfulness of opaque smoke and gases render it necessary to limit the diffusion of gas and smoke. This problem has already been dealt with in many national regulations, at least as regards buildings particularly susceptible to fire (high-rise buildings, department stores, etc.).

30. NKB, CMEA and EEC have not included this problem in their programme. On the other hand, ISO is studying the standardization of smokeproof doors but has not yet fixed a date for the submission of a draft. Nor has it included in its programme the advance calculation of smoke movements. The Building Research Institute of Japan has, however, developed methods of forecasting the diffusion of gas and smoke by calculations.

31. It may be noted that Commission CIB/W 14 is trying to prepare a report as a basis for a possible international code.

METHODS OF TESTING FIRE RESISTANCE

32. There is an International Standard, ISO 834, first edition 1975-11-01. NKB follows this Standard 834. So does CMEA, for the moment at any rate, for the provisions on fire resistance are now being revised and certain minor divergences from ISO 834 may well emerge. A recommendation that CMEA should follow the ISO standard to the letter is obviously in order. EEC is preparing a directive which, while not specifically referring to the ISO standard, reproduces its contents *de facto* and gives certain details of the test and of its interpretation. This directive might be referred to the Council for a decision during 1976.

33. It appears therefore that there is no major problem on this point, and that the Working Party on the Building Industry should recommend that :

> Regional organizations and Governments should include a specific reference to ISO 834 in their regulations, and
>
> ISO should supplement ISO 834 as soon as possible with the necessary details concerning implementation and interpretation.

METHOD OF TESTING INFLAMMABILITY, THE SPREAD OF FIRE AND COMBUSTIBILITY

34. It is a well known fact that national inflammability and spread tests, which are often combined in reaction tests, vary from country to country, and that this is a major obstacle to trade in building materials and components. It is therefore particularly important that the sub-regional organizations should use identical tests.

35. ISO work is making good progress in the fire reaction field (i.e. inflammability and spread) but not in the case of combustibility, as the draft ISO/DIS 1182 has just been rejected.

36. EEC is active in the reaction field. Test methods for the Nordic countries are being worked out in co-operation with a special organ, Nordtest, which receives financial support from the Nordic Council of Ministers. The test methods are being worked out in consultation with NKB in accordance with a special consultation agreement and have, as far as possible, been adjusted to the methods elaborated by the international organizations. CMEA is giving top priority to combustibility - on the basis of ISO/DIS 1182 - and has accorded the other topics second priority.

37. In the remainder of this report, a distinction - based on the ISO approach - will be made between the following :

Inflammability,

Spread, and

Combustibility.

This division is not accepted in all regulations, and it would be useful if the Working Party on the Building Industry could try to persuade all countries to adopt the same concepts.

38. It would, however, be quite feasible, as a means of determining fire reaction (and also, as will be seen later, toxicity), to use the semi-full-scale test. This test, which, despite its innumerable variants, is more realistic than any test on a small sample, has so far neither been standardized nor used in the regulations of any country. For this reason, it would probably be easier to adopt it than to standardize the tests currently in use. To facilitate matters even further, it might be recommended that this test should simply be used side by side with national small-scale tests until the latter had been harmonized or standardized by the means discussed below. The Working Party on the Building Industry could stimulate thinking on this subject.

INFLAMMABILITY TESTS

39. These tests are intended to evaluate materials from the point of view of how they catch fire. ISO/TC 92 has a working group, namely WG 4 "Reaction of Fire Tests", which is to submit a preliminary draft test to the forthcoming plenary session of TC 92, at the end of 1977. EEC has sent a questionnaire on reaction to fire to the nine Governments. Processing of the replies began in March 1976, and it is hoped that a draft can be submitted by the end of 1977. CMEA has no high-priority projects on the subject. The most appropriate

procedure, therefore, would be to recommend that ISO should maintain its proposed time-table and that EEC should base its future directive on the ISO draft, which could also be adopted subsequently by the other regional organizations.

SPREAD TESTS

40. These tests are designed to determine how the flame spreads over a material or a component. They entail similar work and are of the same priority as the inflammability tests with which they are usually grouped under the heading of fire reaction tests. The above recommendations are therefore also valid in this case.

COMBUSTIBILITY TESTS

41. This heading embraces tests which are designed to measure the total number of calories which the combustion of a material can release in various circumstances and the number of calories released over a period of time. ISO/TC 92 has included in its programme the preparation of a test to determine "heat release rates" which should be completed by 1980. It has also been working for a considerable time on the determination of the volatile calorific value by means of the vertical tube test (ISO/DIS 1182), but this project has now come to a halt. Neither EEC nor NKB have any projects in this field at the present time. On the other hand, CMEA was to have submitted a preliminary draft, based on ISO/DIS 1182, of a method of testing the combustibility of materials in June 1976. The ISO draft has, however, been definitely rejected. As the absence of an ISO recommendation is likely to create difficulties, consideration might be given to the possibility of recommending that the regional organization should wait until 1980 before taking any decisions in this field. Otherwise, the Working Party on the Building Industry would have to try to bring the views of these organizations closer together in the hope that a common method can be agreed upon.

METHODS OF TESTING TOXICITY

42. The use of synthetic macro-molecular materials has increased the hazards inherent in the action of carbon monoxide (CO) and has created new hazards (such as an incapacitating effect) caused by gases or mixtures of gases. Consequently, new regulations are about to be issued or have already been published. It is extremely important, if new barriers to trade are to be avoided, that a test should be decided upon as a matter of urgency and accepted throughout the ECE region.

43. However, ISO/TC 92 has been unable to give a date for the submission of a document. A working group on the subject held its first meeting in February 1976. CMEA has not included the topic in its current programme but will begin work on it this year, with 1980 as the target date. EEC (DG 11) has sent a questionnaire to the members of the European Community on the subject of toxicity. The first discussions were to be held in the spring of 1976, and it is anticipated that the technical drafting of a recommendation will be completed by the end of 1977. The text was formulated by the experts but has not yet been accepted by the Ministers. In addition, EEC is at present considering a project on theuse of the semi-full-scale test. It would therefore seem that

the Working Party on the Building Industry could contribute to work in this field, and might propose that the possibility of accepting the semi-full-scale test could be studied.

METHODS OF TESTING SMOKE OPACITY

44. The opacity of smoke, like the toxicity of smoke and gases, has been increased by the use of new materials, whence the need for tests to determine opacity (or, more exactly, the resultant loss of visibility). ISO/TC 92 is preparing an opacity test which should be submitted by the end of 1977. There is, however, a test which is well known and already in use, namely, the NBS test (National Bureau of Standards - United States of America), the adoption of which as an ISO test has been proposed. The proposal has majority support, but if that majority is not sufficiently large, ISO will have to draft a specific text. The subject is not on the EEC programme, but CMEA is to include it in its next programme (1976-1980).

AUTOMATIC ALARM AND FIRE EXTINCTION EQUIPMENT

45. The principles outlined in paragraphs 4 to 9 bring out the importance of automatic alarm and fire extinction equipment. Such equipment must obviously be reliable. It is very important, therefore, that requirements as to its performance should be standardized. The subject is not covered by the programmes of the intergovernmental organizations, nor by those of ISO and CIB as far as automatic fire extinction equipment is concerned. ISO/TC 92 and CEN are, however, examining detectors, but have not drawn up a time-table.

PROPOSED ACTION

46. In the light of the above identified areas of work, the Working Party on the Building Industry may wish :

(a) to recommend to the CIB that high priority be given to the work of its Commission W 14 (see annex I) and, in particular, to work on the draft "Principles and guidelines for international codes on structural fire protection" (see annex II) ;

(b) to recommend to the ISO to give high priority to work on fire testing methods and fire-terminology ;

(c) to request the secretariat to organize an _ad hoc_ meeting of experts from public authorities and international organizations concerned on the subject of "fire protection requirements" for November 1977 to review the progress made in the implementation of the various projects and works recorded in this report, in particular in the fields of terminology (ISO/TC 92), fire test methods (ISO/TC 92), codes on structural fire protection (CIB/W 14) and structural fire protection of timber (CIB/W 18) as well as the activities of the sub-regional intergovernmental organizations (CMEA, EEC, NKB) ;

(d) to request the secretariat to organize an informal consultation with participation of the rapporteurs and representatives of relevant international organizations (CMEA, EEC, NKB, ISO, CIB, etc.) during the spring of 1977 in preparation for the _ad hoc_ meeting (see (c)) ;

(e) to request the secretariat to continue to keep informed of the activities of sub-regional, intergovernmental organizations and non-governmental organizations in the different sectors of the field and ensure that the relevant documentation and information is exchanged between the organizations.

ANNEX I

CURRENT ACTIVITIES OF CIB W 14

(The following list of activities includes those agreed at the last meeting, April 1976)

INTERNATIONAL STRUCTURAL CODES

1. About three years ago, W 14 was asked by various international bodies to take a leading role in harmonizing existing codes. This was effected through the Liaison Committee of International Civil Engineering Organizations in which CIB is formally in liaison with FIP, CEB, CECM, RILEM, IABSE and IASS. It was agreed that the materials associations would produce codes for structural fire safety but that W 14 would review and co-ordinate one with another, primarily with a view to seeing that the codes made similar assumptions about the nature of fire in buildings, did not give vastly different weight to the same information or made different assumptions about the significance of published information. Subsequently, representatives of these organizations agreed that W 14 should produce documents on the general principles of fire safety in buildings to which they, the materials associations, could work and provide solutions for the varying requirements.

2. A W 14 meeting in April 1974 appointed a panel (CAP) to undertake the necessary activities on its behalf to cover the following aspects either singly or together as appropriate.

(a) Scrutinizing of codes before publication.

(b) Advising code committees on latest knowledge.

(c) Adopting a long-term role of producing data on structural fire protection including philosophy.

The Code Advisory Panel (CAP) was constituted as follows :

Mr. H.L. Malhotra (Convener) Fire Research Station, United Kingdom
Colonel A. Cabret - CSTB - France
Professor K. Kordina - Technical College Braunschweig - Germany
Professor O. Pettersson - Lund University - Sweden
Professor J. Witteveen - TNO Delft - Netherlands

3. During the early stages of discussion the special needs of the member countries of the EEC became apparent as a number of activities were being pursued in this field for European directives on structural safety in fire.

To provide assistance in this connexion, Dr. Sfintesco of CECM, who had been retained as a consultant by an EEC Committee was co-opted to the panel. An invitation was also extended to Professor G. Blachère in his capacity as consultant to the ECE secretariat. The Code Advisory Panel has, as a result of discussion, agreed to fulfil its function by undertaking the following activities :

(a) To prepare a document dealing with the nature of the problem and an assessment of fire severity to which structures are likely to be subjected.

(b) To prepare a document on the levels of protection a structure should be designed to provide.

(c) To prepare a document on the principles and guidelines for international codes.

(d) To comment on international codes as and when they are presented.

4. W 14 has interest in the international unification and harmonization of codes dealing with the general principles as well as individual materials and is working in the following areas :

(a) Nature of the problem of fire safety in buildings and the relevance of structural fire protection measures.

(b) Levels of protection a structure should be required to provide on the basis of the expected fire severity.

(c) CAP will offer its comments and observations on codes prepared by materials organization CECM, FIP, CEI Bois, etc.

It has already made some comments on the FIP document on the recommendation for the design of reinforced and prestressed concrete structures.

STRUCTURAL DESIGN AND TESTING

5. The first draft of a report "The design of fire exposed structural steel beams and columns" is in preparation and members are commenting on the first draft prepared by Dr. Magnusson (Lund University, Sweden). On the basis of the co-operative research study on furnaces used for fire resistance testing, a recommendation has been made to ISO to incorporate a new clause in furnace test standards standardizing the refractory that forms the internal furnace surface. This will improve the agreement between tests performed according to ISO R834 or standards related to that.

BUILDING FIRE SAFETY

6. A report "Principles of building fire safety" is being produced. Members are submitting titles for a bibliography, after which a new draft incorporating comments will be circulated for final approval and submission to the Board of CIB. The latest draft is being studied for comment by specialist fire groups within FIP, CECM and RILEM, etc.

SMOKE AND TOXIC PRODUCTS

7. A review of the toxic effect of combustion products as causes of injury and death is being prepared by Dr. Berl (Johns Hopkins University, United States) and is expected to be circulated for comment by October 1976. As a result of a public symposium held in 1975 on the control of smoke movement, a W 14 report is to be prepared giving guidance and the basis for a practical design code. Monsieur Ferrie (France), in collaboration with other W 14 members in France, Germany, Netherlands, United States, United Kingdom, Canada, Japan and Australia, is preparing a first draft.

STATISTICS AND REPORTING

8. A report "The requirements for statistical data on fires in buildings" is in draft, and members are submitting existing reporting forms for inclusion in an appendix, after which a new draft incorporating comments will be circulated for final approval and submission to the Board of CIB. A draft report on the reporting of large fires is being prepared by Mr. Sjölin (FOA, Sweden) and will be circulated soon for comment, to members agreeing to assist in compiling lists of large fires.

NEW ACTIVITIES BEGUN IN APRIL 1976

9. (a) Dr. S. Martin (Stanford Research Institute, United States) is currently reviewing information on large-scale enclosure fire experiments. Members of W 14 agreed to assist in this activity so that a comprehensive index of such experiments in the last few years could be prepared.

(b) Mr. Adam (UTI, France) is going to co-ordinate information on forthcoming large-scale enclosure fire experiments and to co-ordinate suggestions from W 14 members for co-operative research employing large-scale facilities.

(c) Professor Olesen (University of Aalborg, Denmark) will be co-ordinating information from members on specialist teaching curricula for fire studies (in particular for civil engineers) with a view to producing CIB recommendations on education in fire technology.

(d) Professor Brady Williamson (University of California, United States) is preparing a review of work on fire behaviour of furnishings.

(e) W 14 agreed with a suggestion by Professor Witteveen (TNO, Netherlands) that it should organize a symposium on the levels of fire protection and how they should be evaluated. It is hoped to hold this symposium in the spring or summer of 1977 to initiate an international discussion on the question of what are the proper requirements for fire protection and the methodologies for evaluating them.

ANNEX II

PRINCIPLES AND GUIDELINES FOR INTERNATIONAL CODES ON STRUCTURAL FIRE PROTECTION

(draft under preparation by CIB W 14)

1. This document has been prepared under the authority of the CIB/ Commission on W 14 as a consequence of the International Engineering Societies' agreement in 1974 by which CIB was made responsible for the co-ordination of code activities for structures to resist the effects of fire. It is the intention that various international organizations when engaged on the preparation of design codes on structures will keep in mind the principles and the aims which have been put down and follow the method for the arrangement of the code. The first two sections need not occupy much space but the background information should be given in some detail. The main part of the codes is expected to deal with sections 4 and 5 on the design of structures and section 6 provides an opportunity to show how the computational rules can be followed. Section 7 is intended to be a quick reference section in which the designer but more particularly the other users will find ready reference to generic systems.

GENERAL PRINCIPLES

2. A rational approach to the design of a building or structure should consider the probability of the occurrence of a fire, its likely severity, the degree of protection that needs to be provided for purposes of life safety as well as reducing property damage, the consequential effects of a fire and the desire of the community to resist certain types of disasters. At present a comprehensive philosophy for this purpose is not available and the building codes make many simplistic assumptions. One such assumption concerns the severity of a fire and it is common to represent it by the time/temperature relationship of the type specified in ISO/834 specification on fire resistance tests. With new knowledge it is now possible to propose other types of temperature/time relationships which may be more realistic for certain situations. At present in the majority of countries the national regulations, by-laws or building codes typify the expected fire hazard in different buildings on the basis of their occupation, size and location into periods of fire resistance expressed in minutes or hours (ranging from 15 minutes to six hours). In one or two countries an alternative approach is permitted which enables the designer to compute the fire resistance on the basis of the expected fire load, the building characteristics which control ventilation and the nature of the compartment boundaries. Most of the existing building authority documents do not make it clear whether the requirement has to be satisfied by the building as a complete unit, or one part of it or each and every recognizable element.

In practice it is the last mentioned concept which finds expression in codes. Each element of a building is individually required to provide a certain resistance to the effects of fire. Whilst these are some of the concepts which the general code should discuss, it is necessary for the individual structural codes to clearly indicate the basis on which that particular code is designed.

3. A code can in its simplest form follow one of the two routes in providing information. It can specify additional safety factors and appropriate material properties to be used when designing a structure to resist the effects of a specified fire. This may be termed as a rational design approach which should be the ultimate aim of all codes but which may not be realized at present. The second approach in its simplistic form consists of tables or diagrams which specify sizes, thickness of components or parts, cover, additional protection which on the basis of empirical relationships are considered capable of resisting the effects of fire for a minimum period. In some instances additional measures may be specified to counter the effects of expansion, deformation and thermal stresses and strains. This can be termed as a dimensional design approach and seems to be followed in many countries at present.

4. With either approach an assumption has to be made of the live loads, regular as well as accidental, which the structure is designed to resist, and the loads which can be assumed to be the most likely in the case of a fire. The principle effect of a fire is to lower the resistance of structures to loads and consequently the amount of this reduction is critical in the estimation of fire resistance.

5. A code should take into account the range of materials with their differing properties which may be used and allow a designer to make a sensible choice. It should concern itself with the standard of workmanship that is likely to be present on site and realize that the sample constructions examined in a laboratory invariably represent the high rather than the average standard for that type of construction. A number of recent incidents have highlighted the omission of materials, poor fabrication, deficient jointing techniques and misplacement of important components.

6. A code should attempt to present the information in the form in which it can be made use of by designers who have little knowledge of the fire problems and who may not wish to spend much time in learning the background. The information can be presented in a number of ways viz :

(a) step by step computational techniques,

(b) diagrams or graphs, and

(c) tabulated data.

7. In the field of structural fire behaviour, there is still a great deal to learn, and any code drafted at this time will of necessity not be able to deal completely and fully with all the issues. It is therefore necessary to point out to the user the areas where further knowledge is needed and further research is in progress, particularly if he is being asked to accept some interim recommendations.

8. A code should give as many graphical illustrations as considered necessary to explain and to demonstrate certain problems which may remain unclear without a long written explanation.

9. The language, terminology and symbols used in a code should be those internationally acceptable and understood. Full use should be made wherever needed of the appropriate ISO specifications and guide notes.

THE AIMS

10. (1) On the basis of the general principles outlined the main objectives of a structural code are considered to be :

(a) information for the design of buildings subject to fire resistance requirements,

(b) data for checking the adequacy of a given design,

(c) a basis for control purposes by the building authorities.

Some of the additional objectives which such a code may have are to

(d) promote rational design of building structures,

(e) facilitate exchange of information and design data between different countries,

(f) assess the re-use of a structure partially damaged by fire.

(2) The first step in the preparation of a structural code is to consider the general principles of fire safety in buildings and the method that might be used for the quantification of fire resistance requirements. Separate documents are being prepared for this purpose and the factors which are being considered include :

(a) Identification and evaluation of the fire hazard

(b) Overall philosophy for protection

(c) Considerations of life safety

(d) Considerations of property safety

(e) Economics of fire costs and fire protection

(f) Building use/type and level of risk.

(3) The structural code should be prepared with the objective of providing engineers with means for designing structural elements and complete structures to withstand fires to varying degrees, to check structural design for fire resistance and to enable a prediction to be made of the likely behaviour of a structure in case of a known fire.

(4) The structural code should provide background information on the properties of materials at high temperatures as far as available which are relevant towards understanding structural behaviour under fire conditions and which may be used in design calculations.

(5) The structural code should give summarized information and references which have been used in the preparation of the data and which can be used by engineers for further studies of the problem.

(6) The structural code should consider the possibility of providing computational techniques as well as directly tabulated data for assessing fire resistance of elements. It should also indicate if it is possible to extrapolate the data to structures of larger and smaller dimensions. The basis for either method should be explained and reservations, if any, indicated.

(7) The structural code should endeavour to examine the assessment of the fire resistance of the building as a whole either on the basis of the expected behaviour of the individual elements or on grounds of other data.

(8) The structural code should consider situations where the re-use of partially damaged buildings could be an economic consideration and provide information on damage assessment techniques and methods to be followed for repairing such damage. These repaired structures should be capable of providing the same degree of fire resistance as the original construction.

(9) The code should indicate areas where uncertainties exist and the possibility of obtaining information by means of ad hoc full-scale tests (semi-natural tests) on a representative form of the construction.

(10) The code should attempt to provide a list of generic systems or products which can provide the required fire resistance, with the list being revised periodically.

THE METHOD

11. The code drafting committee should attempt to frame the codes, so that the following areas are covered and as much information as possible is provided. Figure 1 outlines in a very simple way various steps which are necessary to ensure that a given structure will fulfil its loadbearing function in case of a fire. This is usually achieved by ensuring that the X value is greater than Y. Somewhat similar considerations apply in making an assessment of the separating structures and their ability to prevent the passage of fire or excessive heat, but the basis for such assessment is usually not analytical.

12. Figure 2 shows in slightly greater detail various routes which can be taken to ensure that the desired fire resistance is being provided. The practical levels of protection to be provided in different types of buildings is considered outside the direct interests of a design code and consequently the activities shown above line Z-Z will not be covered by the design code although a knowledge of various basic concepts is necessary to make a full use of the different techniques.

13. The design codes will generally provide information in one or more of the following ways which are listed in the increasing order of complexity :

1. Tabulation of minimum sizes for different types of building elements and their combinations. In some cases this means constructional specifications for different constructional and protection materials to satisfy fire resistance requirements.

2. In addition to 1, simple formulae or graphs to allow some flexibility in the selection of sizes and materials on the basis of data from the standard tests.

3. In addition to 1 and 2 above, an analysis of the structure can be made, the nature and the amount of loading and effect of the different boundary conditions or structural interaction taken into account.

4. A fully analytical approach in which the computed fire severity is expressed as a functional relationship or equated to an equivalent exposure duration for a standard relationship, the performance of the structure is computed by an analysis of heat transfer to its interior and its effect on structural behaviour. Both of these components have sub-systems and finally a comparison is made between the minimum structural strength that is necessary to prevent collapse or some other specified limit state and the minimum strength the structure will be reduced to by the effects of heat.

14. The following list of headings under which information can be provided is neither exclusive nor may be appropriate for some materials but it nevertheless provides a useful framework which the code committees are asked to use as a starting point. This will facilitate intercode comparison and will be of assistance to users of the codes.

LIST OF HEADINGS

1. Introduction

In addition to providing the background information to the preparation of the code the section may also make reference to the principles used in the drafting of the code, the levels of protection used and the relationship, if any, with the safety regulations which will make use of the document.

2. Scope

(a) Types of constructions covered by the code.

(b) Basis for the expression of fire severity

(c) Type of solutions the code is able to provide.

(d) Future trends and developments.

3. Background data

(a) Identification of factors which affect structural behaviour.

(b) Properties of materials and their behaviour at high temperatures.

(c) Heat transfer methodology employed.

(d) Methods of structural analysis used.

4. Design procedures

(a) Individual major elements, i.e. columns, beams, floors and walls.

(b) Effects of continuity and composite action.

(c) Frames and structural systems.

(d) Junctions and supports.

(e) Methods of enhancing performance by protection.

5. Design aids

Any special aids such as nomograms, computer programmes which are available.

6. Design examples

Illustrations for using any complex techniques.

7. References

The references to useful literature on basic concepts, material properties, design techniques, computer programmes and catalogues of instructions.

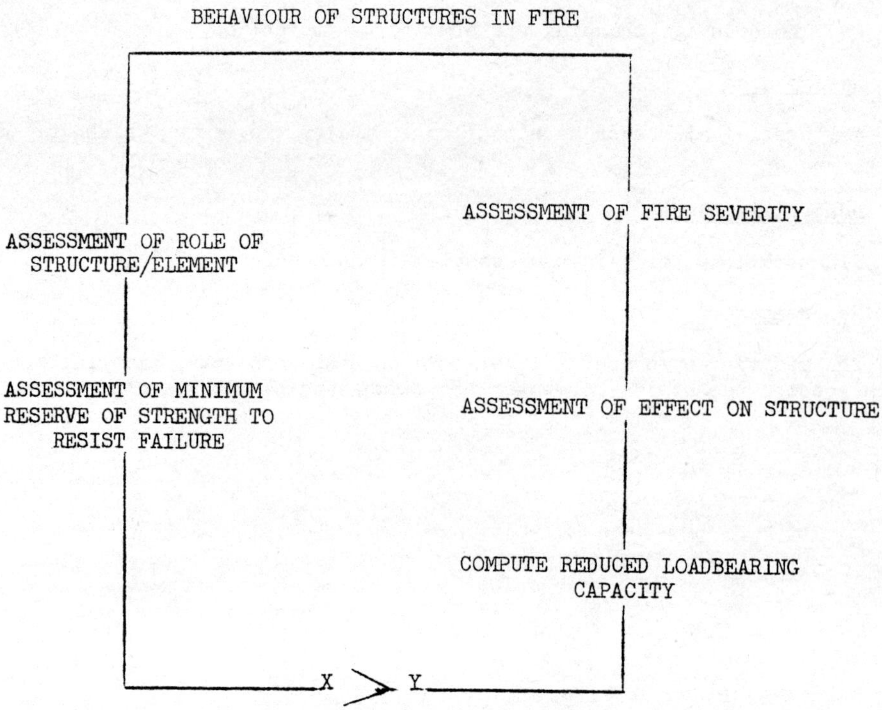

Fig. 1 Different Steps in Assessment of Structural Adequacy.

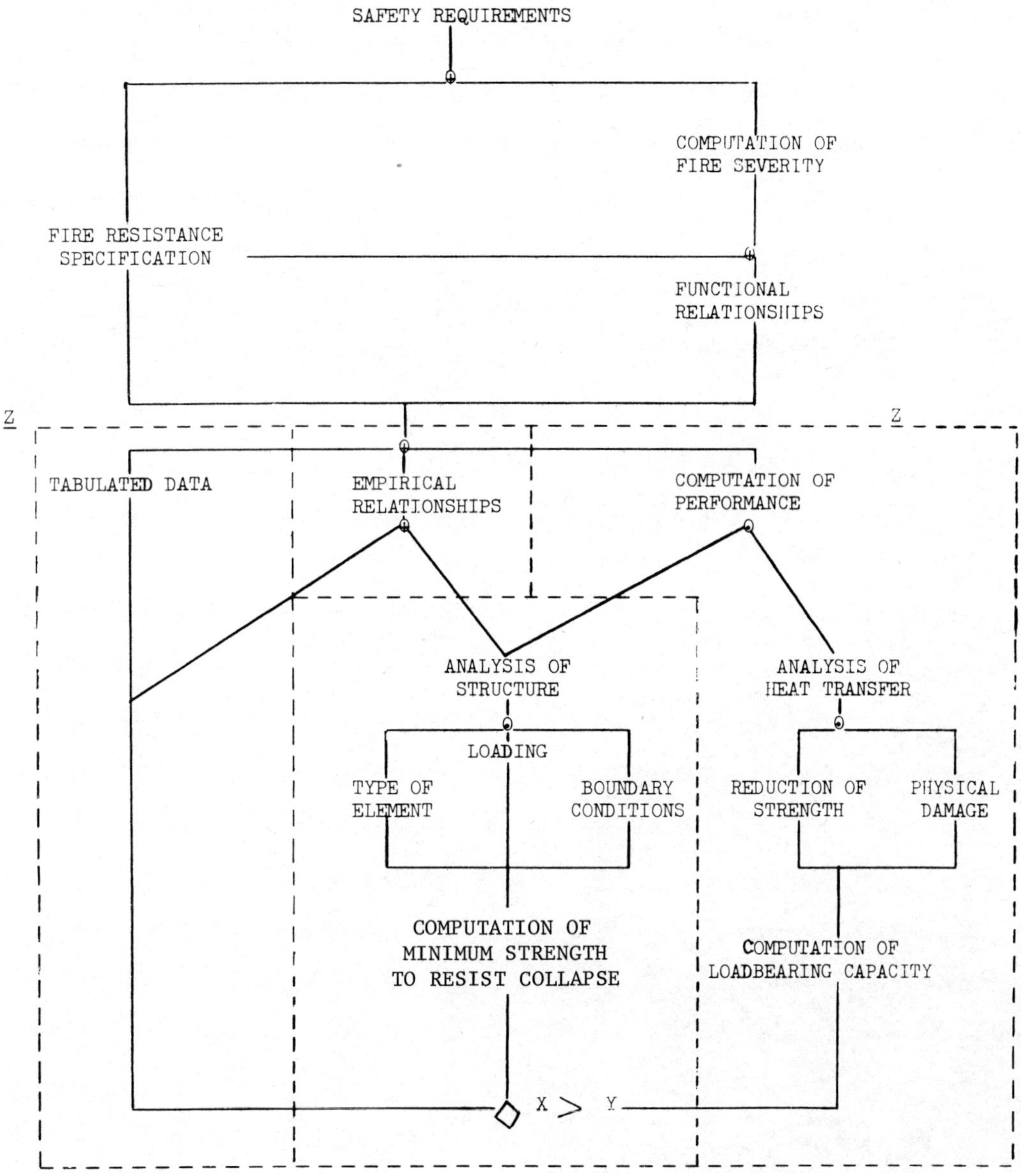

Fig. 2 Means of judging structural resistance to fire.

Notes X > Y (X to be greater than Y) — All paths apply; Alternative paths

Activities on International Harmonization of Building Codes and Regulations

H.L. Malhotra BSc (Eng), MICE, FI Fire E
Building Research Establishment, Fire Research Station
United Kingdom

Amongst various harmonization activities in the building field now in progress, CIB is playing a leading role in the fire field. By mutual agreement the International Association of Civil Engineering Associations (CEB, CIB, ECCS, FIP, IABSE, IASS, RILEM) CIB was asked to co-ordinate the activity on fire matters and it made one of its Commissions (W14) responsible for this work. Commission 14 set up a special panel, the Code Advisory Panel (CAP) with a core membership of five experts from different fields to examine the problem and prepare a statement on work to be done and actions to be promoted. The panel at its earlier discussions found it necessary to co-opt or invite members from other international and European bodies to participate in its discussions. It was fortunate to have the assistance of G. Blachere, consultant to the ECE committee on housing and Dr. Deppe as a representative of CEI Bois.

It saw its interests spanning the whole spectrum of material interests and its activities of relevance to the ECE as well as EEC work. Consequently, it has prepared a strategy along the following lines :

1) To assist in the formulation of basic fire concepts for better understanding of the problems and encouraging a rational approach,

2) To prepare an analytical approach to the design of fire protection measures in buildings,

3) To generate discussions on the principles to be used in determining the level of protection buildings need,

4) To prepare guidelines for structural codes for fire protection, and,

5) To assist various bodies preparing such codes with the problems of harmonization.

Its activities to date are described in some detail in the annex to the ECE document HBP/WP.12/R.74 - October 1976, and its own reports to its parent organization. At present it is assisting with drafts (items 1 and 2), with the CIB seminar on levels of protection in buildings (item 3), agreed guidelines on structural codes (item 4) and is getting ready to assist with item 5 although its individual members are already engaged on that work.

It may be of some interest to list the various activities which are progressing on the harmonization front in different international organizations :

CEI Bois	:	Preparing the initial draft on the behaviour of timber structures.
ECCS	:	Finalizing its draft on the fire behaviour of steel structures.
FIP	:	Acting jointly with CEB to update its design guide for concrete structures on the lines of the CAP proposals.
CIB/W23A	:	Making a start on masonry structures.
ISO	:	Preparing international specifications for fire test standards.
CEN	:	Preparing specifications for detectors, extinguishers and sprinklers.
RILEM	:	Setting up a new committee on material properties at high temperatures representative of the fire environment.
CIB/W14	:	Studying basic fire science and evolving a technical approach to safety.
EEC	:	Preparing draft directives on fire test standards, and considering harmonization of structural codes and building regulations with the aim of removing trade barriers amongst member countries.
ECE	:	Preparing guides for the harmonization of the technical content of building regulations.

This brief description of activities indicates that work is in progress in almost all of the relevant aspects of fire protection and fire behaviour but because of the different interests and different outlooks of various organizations the need for co-ordination will, for some time to come, remain. The Code Advisory Panel is trying to facilitate such co-ordination with its activities.

Annexes

Recommendations and Future Action

The seminar prepared the following statement of its views and recommendations for consideration at the thirty-fifth session of the ECE Timber Committee in Geneva in October, 1977.

Recommendations and Future Action

The seminar concluded that in most countries there was no direct discrimination against wood products, as such, in national building codes. A number of exceptions were cited, however, Hidden discrimination against products termed combustible often occurred, and the degree of such discrimination differed widely from country to country. Efforts towards the harmonization of building codes, such as those being made by ECE, should help to reduce or eliminate many of these anomalies.

The seminar welcomed the intention of ECE to organize an *ad hoc* meeting of experts on the subject of fire protection requirements in November 1977, and recommended that the Timber Committee draw the results of the seminar to the meeting's attention. In particular, it emphasized the need to take a comprehensive look at the problem of fire protection, so that the more detailed work being undertaken in specific areas could be related to an overall objective.

The seminar recommended that, taking account of the work already performed by the International Council for Building Research, Studies and Documentation (CIB), member governments be encouraged to improve and standardize the methods of recording and analyzing the information obtained from routine reports on fires in buildings. The seminar expressed concern at the lack of uniformity in statistics on fires in buildings and the consequent difficulty of drawing overall conclusions and comparisons regarding fire incidents. The seminar recommended that the ECE Committee on Housing, Building and Planning should collaborate with CIB on the recording and reporting of such statistics.

There was a need for co-ordination at the international level of research on the behaviour of wood products in fire. The seminar recommended that the Timber Committee draw this need to the attention of the International Union of Forest Research Organizations (IUFRO), which appeared to be the body best placed to carry out this activity with the purpose of reducing duplication of effort and allowing a faster and fuller exchange of information.

The seminar recommended that CIB, in cooperation with other competent bodies, be asked to accelerate the development of predictive design methods for structural stability of structures in fire. These should be supported by information from tests on materials and assemblies, and their fire performance.

The seminar recommended that timber interests should have more direct access to the deliberations of the TC 92 Committee of the International Organisation for Standardisation (ISO) on fire tests for building materials. This could be achieved by organizations such as CIB and the Confederation of European Timber Industries approaching ISO TC 92 with offers to assist in this work.

The seminar recommended that member governments be strongly urged to implement those fire test methods already published by ISO TC 92 and to encourage, by provision of adequate resources for the necessary development, proving and correlation, earlier completion of methods of test currently undergoing development. The seminar recognized that ISO was the only appropriate body to produce harmonized tests but expressed concern at the apparently slow rate of progress in some fields. Additionally, ISO TC 92 should be encouraged to consider the relative importance of contents (e.g. furnishings) versus building structure components as determinants of fire development and severity, with particular attention to be paid to the toxicity and associated aspects of combustion.

The seminar took note of the widespread use of untreated wood products as linings for walls and ceilings of single-family dwellings in the Nordic countries. This practice was in accordance with the provisions of the safety building regulations of these countries and without any adverse effect on fire safety performance. The seminar suggested that other countries in the ECE region could benefit from this experience by examining whether their own building codes could be more closely aligned with those in the Nordic countries in this respect.

The seminar deprecated the concept of "non-combustibility" in codes and regulations to define and characterise fire behaviour. It further noted the lack of uniformity in defining "non-combustibility" as practised in various countries. The definition and the test were applied to particular materials and might have little relevance in the context of whole building design. The safety function intended by this criterion could be fulfilled by other more appropriate reactions to fire test procedures.

The seminar stressed that, when considering the role of flame retardant treatments for timber and wood-based materials, a balanced view of their influence on cost and the properties of the treated materials and all aspects of their fire behaviour be taken into account in order to make rational decisions on their appropriateness to particular building situations. Where there were restrictions against the use of wood products, flame retardants had a positive role to play in increasing the use of these materials.

Listing of Research Organizations Concerned with the Behaviour of Wood in Fire[1]

Timber Research and Development Association, United Kingdom

Australia	C.S.I.R.O., Division of Building Research Graham Road, Highett Melbourne, Victoria 3190
Austria	Magistratsabteilung 39 Eslarngasse 23, A-1030 Vienna
	Versuchsanstalt für Material-prüfung Petzoldstr. 45/47, A-4020 Linz
Belgium	University of Gent St Pietersnieuwstraat, Gent
Canada	Canadian Wood Council 77 Metcalf Street, Ottawa
	Eastern Forest Products Laboratory 800 Montreal Road, Ottawa KIA OWS
Czechoslovakia	State Forest Products Research Institute 5 Lamacska cesta, Bratislava
Denmark	National Institute for Testing of Materials Amager Boulevard 108, 2300 Kobenhavn S
	Laboratory of Heating and Air Conditioning Technical University of Denmark DK 2800 Lyngby
	The Danish Fire Research and Development Group Parkalle 345, Copenhagen 2600 Glostrup
	Danish Society of Fire Technology Ingeniørhuset, Vester Farimagsgade 31 1606 Copenhagen V
	Institute of Building Technology and Structural Engineering Aalborg University Centre Danmarksgade 19 9000 Aalborg

1/ This listing was presented as a draft to the Seminar and was revised and added to by the participants. It is not intended to be comprehensive.

Denmark (continued)	Danish Fire Protection Association Nygaards Plads 9 Copenhagen, 2610 Rødovre
Finland	Technical Research Centre of Finland (Valtion Teknillinen Tutkimuskestus) Fire Technical Laboratory 02150 Espoo 15
France	Centre Technique du Bois 10 Avenue de Saint Mande Paris 12e
	Centre Scientific et Technique du Bâtiment 4 Avenue du Recteur Poincare Paris 16e
Germany, Federal Republic of	Institut für Holzforschung und Holztechnik University of Munich Winzerestrasse 45 D-8000 Munchen
	Institut für Baustoffkunde und Stahlbetonbau Beethavenstrasse 52 33, Braunschweig
	Bundesanstalt für Materialprüfung Unter den Eichen 87 D-1000 Berlin 45
	FMPA Bauwesen Otto-Graf Institut Pfaffenwaldring 4 D-7000 Stuttgart
Hungary	Institute for Testing Building Materials and Structures Dioszegi Ut 37 Budapest XI
Japan	Building Research Institute 4-Chome Hyakunin-cho Shinjuku-ku Tokyo
	Fire Research Institute 14-1 Nakahara 3-Chome Mitaka Tokyo
Netherlands	Institute TNO for Building Materials and Structures 5 Lange Kleiweg Rijswijk

New Zealand	Building Research Association of New Zealand 42 Vivian Street P O Box 8375 Wellington
Poland	Wood Technology Institute Poland Poznan, ul Winiarsha 1
	Building Techniks Institute Poland Warszawa ul Filtrowa 1
Sweden	National Institute of Materials Testing Brinellgaten 4 Box 857 S501, 15 Boras
	Swedish Forest Products Research Laboratory Box 5604 S-114 86 Stockholm
	Technical University of Lund Structural Engineering Department Lund
	Technical University of Stockholm Building Materials Department Stockholm
	Research Laboratory of Defence (FOA) Ursvik Sundbyberg
Switzerland	Institut für Holzforschung, ETH Universitätsstrasse 2 8092 Zurich
United Kingdom	Building Research Establishment Fire Research Station Melrose Avenue Borehamwood Herts.
	Timber Research and Development Association Stocking Lane Hughenden Valley High Wycombe HP14 4ND Bucks.
	Safety in Mines Research Establishment off Broad Lane Sheffield S3 7HQ

USA

US Department of Agriculture
Forest Products Laboratory
Madison
Wisconsin

National Forest Products Association
1619 Massachusetts Avenue
Washington DC
(Fire Research Associate Program at US Dept. of Commerce, National Bureau of Sts.)

Weyerhauser Corporation
Longview
Washington

USSR

VNIIDREV
Moscow
Gritsevets St 2/16